William Murrell

Nitro-Glycerine as a Remedy for Angina Pectoris

William Murrell

Nitro-Glycerine as a Remedy for Angina Pectoris

ISBN/EAN: 9783337340407

Printed in Europe, USA, Canada, Australia, Japan

Cover: Foto ©berggeist007 / pixelio.de

More available books at **www.hansebooks.com**

NITRO-GLYCERINE AS A REMEDY

FOR

ANGINA PECTORIS

BY

WILLIAM MURRELL, M.D., M.R.C.P.

LECTURER ON MATERIA MEDICA AND THERAPEUTICS AT THE WESTMINSTER
HOSPITAL; SENIOR ASSISTANT PHYSICIAN TO THE ROYAL HOSPITAL
FOR DISEASES OF THE CHEST.

LONDON

H. K. LEWIS, 136 GOWER STREET, W.C.

1882

PREFACE.

THE object of this work is to give directions for the administration of Nitro-Glycerine as a remedy for Angina Pectoris, the principal points being illustrated by reference to cases that have been under my care. Some of these cases were published in the *Lancet* in 1879.

WILLIAM MURRELL.

38 Weymouth Street,
      Portland Place, W.
            February, 1882.

NITRO-GLYCERINE AS A REMEDY FOR ANGINA PECTORIS.

NITRO-GLYCERINE was discovered in 1847 by Sobrero, and its chemical properties have been investigated by Railton, De Vrij, De La Rue and Müller, Mills, Dupré, Martindale and others. It is commonly prepared by what is known as Liebe's process. Half an ounce of dehydrated glycerine is poured with constant stirring into a mixture of two ounces of oil of vitriol and one ounce of fuming nitric acid of specific gravity 1·52—the temperature of the mixture being kept below 25° C. (77° F.) by external cooling with ice, and as soon as oil drops begin to form on the surface, the mixture is poured with constant stirring into fifty ounces of cold water. Nitro-glycerine then separates and may be purified by washing and drying in small quantities in a vapour bath. The manufacture is carried on chiefly in Scotland, but there is reason to believe that small quantities are made even in London. The greatest possible care must be taken in the process, the glycerine being added drop by drop and the temperature carefully noted by means of the thermometer. The nitro-glycerine thus obtained must be well washed to free it from traces of acidity. It first appears as a white opaque milky-looking oily fluid, but on careful

B

drying by exposing it in a warm room in flat dishes containing thin layers, it becomes dehydrated and forms a transparent colourless oily fluid. It is slightly soluble in water and freely in alcohol and ether, and it has recently been found to dissolve readily in fats and oils. It is, although slightly volatile, inodorous and has a sweet pungent aromatic taste. It crystallises or freezes at low temperatures. It is largely employed as an explosive in mining and blasting operations, being fired by percussion, and forms the basis of the compounds known as " dynamite," " glyoxylon," "dualin," &c. When boiled with potash it decomposes, glycerine and nitrate of potash being formed.

The action of nitro-glycerine on the lower animals has been investigated both in this country and on the continent. A dose of six minims of a ten per cent. solution, injected under the skin of a frog produced, among other symptoms, languor, tetanus and finally paralysis. Immediately after the injection the animal became restless and the respirations very rapid. In a minute or two this restlessness subsided and gave place to lethargy, the frog showing a disinclination to move. The respiration continued rapid, and in about five minutes from the commencement of the observation the animal gave a sudden spring and fell into tetanic convulsions. These lasted about half a minute and then passed off; they soon returned however and were readily excited by touching the animal. After continuing for some time they gradually became weaker and the animal died. In some instances, the mouth seemed to be the part first affected by the convulsions, as the jaws were seen to open and shut, but possibly this action was connected with respiration, rather than with the general convulsions; next it spread to the fore limbs and finally to the hind ones. It was noticed too that the fore limbs were more sensitive than the hind, as slight spasmodic twitches could sometimes be produced by touching or pinching the former, when similar irritation of the latter had no effect. To ascertain whether the tetanus was due to the action of the drug on the spinal cord, or on the nervous centres within the encephalon, the spinal cord was cut across before the poison was given, the upper part of the animal immediately became very restless, the fore limbs were outstretched with the toes spread out; but there was no alteration in the hinder part of the body or in the hind limbs. This result was confirmed by another experiment. A frog was decapitated, and after the spinal cord had recovered from the shock, and reflex movements were again observed, the drug was injected, but no spasm occurred. Dr. Lauder Brunton thinks that the tetanus is not due to any action on the cerebral lobes, but probably to the effect of the poison on the optic lobes.

The principal effects produced on cats by nitro-glycerine are great acceleration of respiration, paralysis, loss of reflex action and sensibility and death from arrest of respiration. It is recorded that a dose of about sixty minims of a ten per cent. solution having been injected into the peritoneal cavity of a cat, there was observed in a few minutes a stretching movement of the hind leg, as if the animal were trying to shake something off the foot. Half an hour later the animal vomited, and at the expiration of about an hour, during which the legs seemed to fail in walking, it sank down never to rise again. Vomiting occurred again once or twice, breathing became very rapid and the tongue was drawn backwards and forwards as in a dog that has been running. Slight spasm-like hiccup then set in, and five minutes later the animal was dead—a little over two hours after the injection of the poison. The loss of reflex action noticed in the observations on the frog and cat, in the advanced stages of poisoning, would indicate that the cord is paralysed; and from the persistence of reflex action, in parts supplied by the cranial nerves, after its disappearance from other parts of the body, it would seem that the spinal cord is paralysed, before the ganglia at the base of the brain.

Some five-and-twenty years ago, a controversy arose as to the properties, physiological and therapeutic, of this substance. The discussion was opened by Mr. A. G. Field, then of Brighton, who described in detail the symptoms he had experienced from taking two drops of a one-per-cent. solution of nitro-glycerine in alcohol. About three minutes after the dose had been placed on his tongue, he noticed a sensation of fulness in both sides of the neck, succeeded by nausea. For a moment or two there was a little mental confusion, accompanied by a loud rushing noise in the ears, like steam passing out of a tea-kettle. He experienced a feeling of constriction around the lower part of the neck, his forehead was wet with perspiration, and he yawned frequently. These sensations were succeeded by slight headache, and a dull, heavy pain in the stomach, with a decided feeling of sickness, though without any apprehension that it would amount to vomiting. He felt languid and disinclined for exertion, either mental or physical. This condition lasted for half an hour, with the exception of the headache, which continued till the next morning. These symptoms Mr. Field describes as resulting from a single dose of one-fiftieth of a grain. Thinking that possibly he might be unusually susceptible to the drug, he induced a friend to take a dose. The gentleman experienced such decided effects from merely touching his tongue with the cork of the bottle containing the nitro-glycerine solution, that he refused to have anything more to do with it. A lady, suffering from toothache, on whose tongue Mr. Field placed about half a drop of the same solution, experienced a pulsation in the neck, fulness in the head, throbbing in the temples, and slight nausea. The toothache subsided, and she became partly insensible, disliking much to be roused. When fully sensible, she had a headache, but the toothache was gone. Another of Mr. Field's patients, a stout, healthy young woman, accidentally swallowed a piece of lint dipped in nitro-glycerine, whilst being applied to a decayed tooth. In about five minutes, after feeling giddy and sick, with headache, she became insensible. Her countenance, naturally florid, was unaltered, breathing tranquil, pulse full, and rather quickened. She recovered in about three minutes, after the administration of a stimulant. Some headache was complained of, but the toothache was gone. Mr. Field, in conclusion, offered some suggestions as to the therapeutic uses of the drug, and stated that he had not met with a single well-defined case of neuralgia or spasmodic disease, in which it had failed to afford some relief.

This paper was followed by a letter from Dr. Thorowgood, in the main confirmatory of Mr. Field's observations. He, after taking a small dose, "experienced a tensive headache over the eyes and nose, extending also behind the ears, and soon followed by a tight, choking feeling about the throat, like strangulation. Neither loss of consciousness nor nausea was experienced, and a walk by the sea soon did away with the unpleasant feeling."

These statements did not long remain unchallenged, their accuracy being called in question by Dr. George Harley, of University College, and Dr. Fuller, of St. George's. Dr. Harley, having obtained some nitroglycerine of the same strength as Mr. Field's, commenced his observations by touching his tongue with the cork of the bottle containing the solution. He experienced "a kind of sweet and burning sensation, and soon after a sense of fulness in the head, and slight tightness about the throat, without, however, any nausea or faintness." After waiting a minute or two these effects went off, and Dr. Harley was inclined to think "they were partially due to imagination." Determined, however, as he says, to give the drug a fair chance, he swallowed five drops more, and as this did not cause any increased uneasiness, he took, in the course of a few minutes, another ten drops of the solution. Being at the time alone, he became alarmed lest he should have taken an overdose, and very soon his pulse rose to above 100 in a minute. The fulness in the head and constriction in the throat were, he thought, more marked than after the smaller dose. In a minute or two the pulse fell to 90, but the fulness in the head lasted some time, and was followed by a slight headache. To two medical friends, Dr. Harley administered respectively twenty-eight and thirty-eight drops in divided doses without the production of any symptoms. Some pure nitro-glycerine was then obtained, and of this Dr. Harley took, in the course of a few minutes, a drop, equivalent to a hundred drops of the solution previously employed. The only symptoms produced were a quickened pulse, fulness in the head, and some tightness in the throat; but as these passed off in a few minutes, Dr. Harley considered that they were probably the effects of "fear and imagination." On a subsequent occasion he took, in the course of three-quarters of an hour, a quantity of the nitro-glycerine solution equivalent to 199½ drops of the solution used by Mr. Field, with the production of no more disagreeable symptoms than those he had experienced in his former trials. The quickening of the heart's action he ascribed to fear, but the head and neck sensations were, he considered, "too constant to be attributed to the same cause," although he thought they were exaggerated by the imagination. Dr. Harley, in conclusion, states that he experimented on ten different gentlemen with nitro-glycerine solution, obtained from four different sources, without witnessing any dangerous effects when administered in the above doses; but he adds that, if taken pure, great caution should be used.

Dr. Fuller, whose observations were made in conjunction with Dr. Harley, commenced by taking two drops of a one per cent. solution. In the course of a minute he felt, or "fancied he felt," some fulness in the head, but was not conscious of any other unusual sensation. A little later he took one-sixth of a drop of pure nitro-glycerine, equivalent to about seventeen drops of the solution spoken of by Mr. Field. Two minutes later his pulse had risen to 96, and there was an increased fulness about the head, but without giddiness or confusion of thought. The pupils were not affected, and he did not experience any unusual sensation beyond that already mentioned. A quarter of an hour later he took a dose equal to 33⅓ drops of Mr. Field's solution, and a few minutes later another dose equivalent to 50 drops. He felt somewhat nervous, and for a few minutes the surface of the body was covered with a clammy perspiration; his pulse intermitted occasionally, and he experienced an increase of fulness about the head. Whether the acceleration of the pulse observed in the first instance was attributable to the effects of the drug he was unable to decide, but his own impression was that it was merely the result of nervousness and excitement; for, had it been otherwise, it is not likely, he says, "that the pulse would have fallen to its natural standard within so short a period after taking the larger doses." The fulness in the head might, he considered, have been attributed in part to the same cause, but a sense of discomfort in the head lasting some hours was, he thought, really due to the drug. As the result of these observations, Dr. Fuller concluded that nitro-glycerine was incapable of producing the effects that had been ascribed to it, and that it might be taken with impunity in considerable quantity.

In a subsequent communication, Mr. Field reasserted the correctness of his observations, and maintained that a reasonable explanation of the very different results obtained by different observers might be found in the great variation in strength to which this drug is liable. He considered, too, that the conditions under which the drug was taken had much to do with its action. When the system is worn out by fatigue, he says, it is more likely to act powerfully than when taken under less unfavourable conditions. On the occasion of taking the dose which produced in him such startling effects, his nervous energy had been impaired by an unusually hard day's work. He found that under more favourable conditions he could take the same dose with the production of nothing worse than headache. Having in his experiments on himself experienced the greatest variation in the strength of different specimens of nitro-glycerine, he was disposed to think, on reading the account given by Dr. Fuller and Dr. Harvey, that they had used a less powerful agent. He accordingly called on Dr. Fuller, and induced him to take a dose of the solution he had used, but to his surprise he experienced little beyond headache. On the same day, Mr. Field administered to a hospital patient suffering from hemicrania two drops of the solution. In about a minute he became pallid, felt sick and giddy, his forehead was covered with perspiration, and he sank on the bed by which he was standing almost unconscious, his pulse failing so as scarcely to be felt. After the administration of a little ammonia the circulation became more vigorous, and in twenty minutes there was a marked diminution of the pain, and he experienced a great desire to sleep, a luxury of which his sufferings had almost deprived him on previous nights. Mr. Field administered small doses of the drug to several other people, all of whom were distinctly affected by it.

Mr. Field's observations respecting the activity of the drug were also confirmed by Mr. F. Augustus James, a student of University College. He took a single drop of the one per cent. solution. In the course of a few minutes he experienced a sensation as if he were intoxicated. This was quickly followed by a dull aching pain at the back of the head, which was alternately better and worse, each accession becoming more and more severe. It soon extended to the forehead and the back of the neck, in which there was a decided feeling of stiffness. He also experienced some difficulty of deglutition, succeeded by nausea, retching, and flatulence. A profuse perspiration ensued, and in a quarter of an hour the symptoms began to abate, but he continued dull and heavy. His pulse, he found, had risen from 80 to 100. Considerable headache remained, which increased in the after part of the day, so that at six o'clock he was compelled to go to bed. At break of day he was not relieved, but after a few hours' more sleep he felt quite well again.

Dr. G. S. Brady, of Sunderland, obtained very decided results from the administration of large doses of nitro-glycerine to a lady suffering from severe facial neuralgia. He gave two minims and a half of Morson's five per cent. solution in a little water. In the course of two or three minutes she began to complain of sickness and faintness; these rapidly increased; there was for a few minutes unconsciousness, accompanied by convulsive action of the muscles of the face, and stertorous breathing. After swallowing some brandy-and-water, she vomited, and the unpleasant symptoms gradually subsided. Dr. Brady also mentions the case of a relative of his, a chemist, who took a drop of the five per cent. solution in water. Shortly afterwards a feeling of sickness and pain at the epigastrium came on, and he left his desk to pace about the shop, thinking to walk off the uncomfortable sensations. Instead of this they grew worse, and an intolerable sense of oppression and swimming in the head, with spasmodic twitching of the limbs, supervened. He had barely time to call his assistant when he fell back insensible. Cold water was freely dashed over the face, and the unconsciousness soon passed away. No vomiting ensued, but the sensation of sickness lasted for some time.

Some four or five years ago being interested in this curious controversy, and quite at a loss to reconcile the conflicting statements of the different observers, or arrive at any conclusion respecting the properties of the drug, I determined to try its action on myself. Accordingly I obtained some one per cent. solution. One afternoon, whilst seeing out-patients, I remembered that I had the bottle in my pocket. Wishing to taste it, I applied the moistened cork to my tongue, and a moment after, a patient coming in, I had forgotten all about it. Not for long, however, for I had not asked my patient half a dozen questions before I experienced a violent pulsation in my head, and Mr. Field's observations rose considerably in my estimation. The pulsation rapidly increased, and soon became so severe that each beat of the heart seemed to shake my whole body. I regretted that I had not taken a more opportune moment of trying my experiments, and was afraid the patient would notice my distress, and think that I was either ill or intoxicated. I was quite unable to continue my questions, and it was as much as I could do to tell him to go behind the screen and undress, so that his chest might be examined. Being temporarily free from observation, I took my pulse, and found that it was much fuller than natural, and considerably over 100. The pulsation was tremendous, and I could feel the beating to the very tips of my fingers. The pen I was holding was violently jerked with every beat of the heart. There was a most distressing sensation of fulness all over the body, and I felt as if I had been running violently. I remained quite quiet for four or five minutes, and the most distressing symptoms gradually subsided. I then rose to examine the patient, but the exertion of walking across the room intensified the pulsation. I hardly felt steady enough to perform percussion, and determined to confine my attention to auscultation. The act of bending down to listen caused such an intense beating in my head that it was almost

unbearable, and each beat of the heart seemed to me
to shake not only my head, but the patient's body too.
On resuming my seat I felt better, and was soon able
to go on with my work, though a splitting headache
remained for the whole afternoon. Were my symptoms due to nervousness or anxiety? Certainly not.
I will not say that I discredited Mr. Field's observations, but after Dr. Harley's positive assertions I certainly did not expect to obtain any very definite results
from so small a dose. Moreover, at the moment of the
onset of the symptoms I was engaged in consideration
of another subject, and had forgotten all about the
nitro-glycerine. I did nothing to intensify the symptoms, but, on the contrary, should have been only too
glad to have got rid of them. The headache, I can
most positively affirm, was anything but fancy. Since
then I have taken the drug some thirty or forty times,
but I never care to do so unless I am quite sure that
I can sit down and remain quiet for a time, if necessary. It uniformly produces in me the same symptoms, but they are comparatively slight if I refrain
from moving about or exertion of any kind. The
acceleration of the pulse is very constant, although
sometimes it amounts to not more than ten beats in
the minute. The temperature remains unaffected.
The pulsation is often so severe as to be acutely
painful. It jerks the whole body so that a book held
in the hand is seen to move quite distinctly at each
beat of the heart. The amount of pulsation may be
roughly measured by holding a looking-glass in the
hand and throwing the reflection into a dark corner

of the room. Before taking the drug the bright spot
may be kept steady, but as soon as the pulsation begins
it is jerked violently from side to side. I have taken
all doses from one minim to ten, sometimes simply
dropped on the tongue, at others swallowed on sugar
or in water. I have not ventured to take more than
fifteen minims in a quarter of an hour. Once or twice
a ten-minim dose has produced less pulsation than I
have experienced at other times from a single drop ;
but then with the larger quantity one is careful to avoid
even the slightest movement. After a five-minim dose
I usually experience a certain amount of drowsiness—
a lazy contented feeling, with a strong disinclination to
do anything.

Thinking there might be individual differences of
susceptibility to the action of nitro-glycerine, I laid my
friends and others under contribution, and induced as
many as possible to give it a trial. I have notes of
thirty-five people to whom I have administered it—
twelve males and twenty-three females; their ages
varying from twelve to fifty-eight. I found they
suffered from much the same symptoms as I did,
although it affects some people much more than others.˙
Of the numbers above quoted, only nine took minim
doses without experiencing decided symptoms. Women, and those below par, are much more susceptible
to its action than are the strong and robust. A delicate young lady, to whom, adopting Mr. Field's
suggestion, I administered it in drop doses for the
relief of neuralgia, experienced very decided effects
from it, each dose producing a violent headache lasting

from half an hour to three hours. A married woman, aged thirty-five, took one minim with very little inconvenience, but was powerfully affected by two. She was obliged to sit down after each dose, and was positively afraid to move. It made her hot, and caused such a beating in her head that she had to support it with her hands. She experienced a heavy weight on the top of the head, and also a sharp darting pain across the forehead, which for a moment or two was very painful to bear. A friend, who for some days took four drops every three or four hours, informs me that at times it affected his head "most strangely." The pulsation was very distressing, and often lasted an hour or more, being intensified by moving. It has relieved him of an old-standing facial neuralgia, and he is enthusiastic in its praise. A young woman, aged 29, complained that after every dose of the medicine—one minim—"it seemed as if the top of her head were being lifted off," and this continued sometimes for five minutes, and sometimes longer. The medicine made her bewildered, and she felt sick. A patient with a faint apex systolic murmur was ordered one minim in half an ounce of water four times a day. He took two doses, but it caused "such a beating, thumping, hot pain" in his head that he was unable to continue it. A young man who was given nitro-glycerine in mistake for phosphorus said it made is temples throb, and he could see his pulse beat so distinctly that he was frightened. It caused a burning and flushing in his face, and "took every bit of strength away." This would last for twenty minutes or half an hour after each dose.

There was no headache. That alarming symptoms may be produced by large doses, is shown by the following case. A woman, aged 51, was ordered drop-doses of the one per cent, solution every four hours. This was taken well, at the expiration of a week the dose was doubled. No complaint being made, it was then increased to four minims, and after a time to six. The patient said "the medicine agreed with her," and even leading questions failed to elicit any complaint of headache or the like. · After the medicine had been taken continuously for five weeks the dose was increased to ten minims. The patient then stated that the medicine no longer agreed with her; it made her sick after every dose and took her appetite away. She always vomited about five minutes after taking the medicine, the vomiting being immediately followed by headache. The medicine made her "go off in a faint," after each dose. She had three "fainting fits," in one day, and could not venture to take another dose. She became quite insensible, and once remained so for ten minutes. Each fainting fit was "followed by cold shivers," which "shook her violently all over." Her husband and friends were greatly alarmed, but she thought on the whole it had done her good. She had never noticed that the medicine produced drowsiness. In another case a three-minim dose taken on an empty stomach caused a feeling of faintness; "everything goes dark," the patient said, "just as if I were going to faint." The patient could take the same dose after meals without the production of any unpleasant symptom. Drowsiness is not an uncommon result of taking nitro-glycerine. A woman who was given drop-doses four times a-day said that she usually went fast asleep immediately after each dose, sleeping from three to four hours. In my own case the desire for sleep was almost irresistible, although the sleep seldom lasted more than an hour. In exceptional cases none of the ordinary symptoms are exhibited. A man with epispadias—to be presently mentioned—took twenty-five minims of the one per cent. solution without any inconvenience.

I have since given nitro-glycerine in some hundreds of cases and the results fully confirm the conclusions at which I had arrived. In some intances it would appear to produce its physiological effect even when merely handled. My colleague Dr. Dupré tells me that he often suffers severely from its effects after examining samples of dynamite. He is always careful to wash his hands thoroughly immediately after touching it, or he soon experiences a most persistent headache, and is sleepless the whole night. It is probable, however, that after a time a certain amount of tolerance is established. Major Majendie, the Hon. Chief Inspector of Explosives, says that he has often been struck by the extraordinary ruddy complexion of the women who work in the dyna-mite factory in Ayrshire and their general appearance of stoutness and health, in fact "a comelier lot of young women it would be difficult to find." It is said, al-though I do not know how true it may be, that many of them suffer from headache only on Monday after they have been away from work for one or two days.

From a consideration of the physiological effects of the drug, and especially from the similarity existing

between its general action and that of nitrite of amyl,
I concluded that it would probably prove of service in
the treatment of angina pectoris, and, I am happy to
say, that this anticipation has been realised.

I was anxious to obtain a comparative series of sphygmographic tracings, and for these I am indebted to the
kindness and courtesy of Dr. Fancourt Barnes, whose
extensive practical acquaintance with the sphygmo-
graph is a guarantee of their accuracy. Dr. Barnes
has taken over 150 tracings of my pulse, some showing the influence of nitro-glycerine, in others of nitrite
of amyl. It would be tedious to describe the observations in detail, more especially as the tracings speak for
themselves, and we consequently give only a summary
of our results. Judged by the sphygmographic tracings, the effects of nitrite of amyl and of nitro-glycerine
on the pulse are similar. Both drugs produce a marked
state of dicrotism, and both accelerate the rapidity of
the heart's action ; they differ, however, in the time
they respectively take to produce these effects. The
full action of the nitro-glycerine is not observed in the
sphygmographic tracings until six or seven minutes
after the dose has been taken. In the case of nitrite of
amyl the effect is obtained in from fifteen to twenty
seconds, after an inhalation, or a dose has been taken
on sugar. The influence of the nitrite of amyl is extremely transitory, a tracing taken a minute and a half
after the exhibition of the drug being perfectly normal.
In fact, the full effect of the nitrite of amyl on the pulse
is not maintained for more than fifteen seconds. The
nitro-glycerine produces its effects much more slowly ;

they last longer, and disappear gradually, the tracing not resuming its normal condition for nearly half an hour. The effect may be maintained for a much longer time by repeating the dose. Nitro-glycerine is more lasting in its power of producing a dicrotic form of pulse-beat, and, consequently, in cases where the conditions of relaxation and dicrotism are desired to be maintained for some space of time, its exhibition is to be preferred to that of nitrite of amyl.

Influence of Nitrite of Amyl on the Pulse.

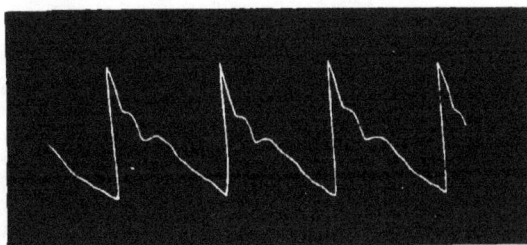

No. 1 —Before inhalation.

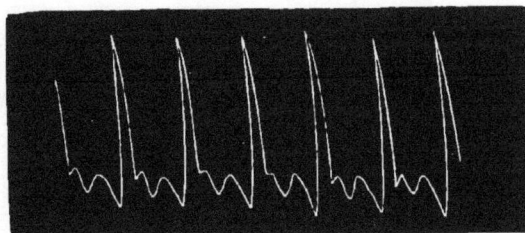

No. 2.—One minute after inhalation.

c 2

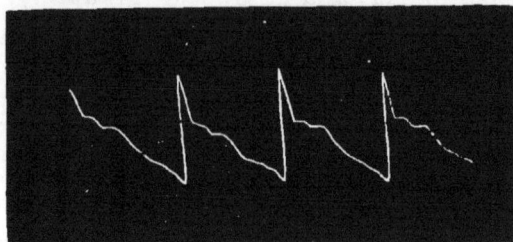

No. 3.—Two minutes after inhalation.

Influence of Nitro-Glycerine on the Pulse.

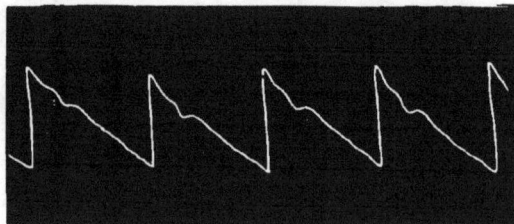

No. 1.—Before dose.

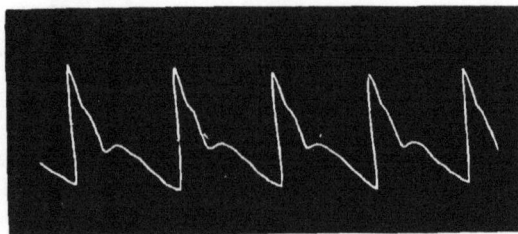

No. 2.—Two minutes after dose.

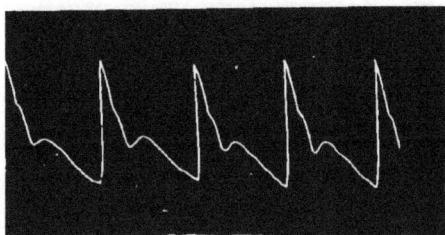

No. 3.—Eight minutes after dose.

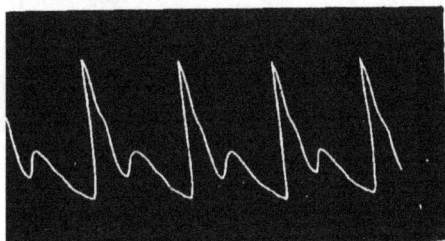

No. 4.—Nine minutes after dose.

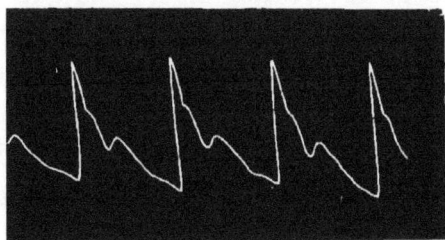

No. 5.—Ten minutes after dose.

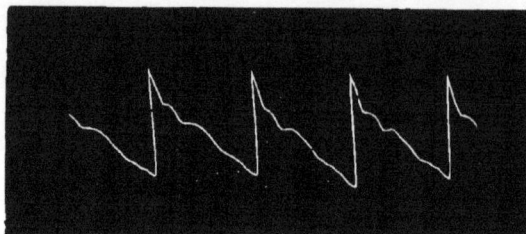

No. 6.—Twenty-two minutes after dose.

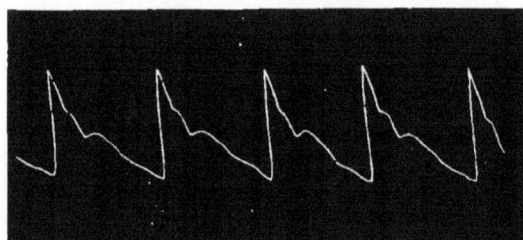

No. 7.—Twenty-six minutes after dose.

Whilst making some observations with nitro-gly-cerine on a patient suffering from epispadias, he called attention to the fact that the administration of the drug always caused an increased flow of urine. On examination, fifty-three minutes after the administration of a dose of twelve minims of the one per cent. solution, the urine was seen spouting from the extremity of each ureter in a little jet some three or four inches high. Ordinarily the urine dribbles away drop by drop, and never spouts out. The patient was much amazed, and said that in the whole course of his life he had never known it go on in that way. If he took beer or spirits it would increase the flow, but this, to use his own expression, "licked everything." He was made to lie on his face, so that all the urine might be collected. In twelve minutes he secreted 6¾ oz. of urine, the specific gravity of which was only 1000. He was then given another dose of fifteen minims in a little water, and in the next twelve minutes he secreted 7¾ oz. Three days later, no nitro-glycerine having been given in the meantime, an observation was made with the view of determining the normal rate of secretion. In half an honr he secreted 3½ oz., the specific gravity of which was 1005. This, he stated, was more than he usually passed, for he had taken three-quarters of a pint of milk about two hours before, and "it was just running through him."

On another occasion a more systematic observation was made. His urine was collected every quarter of an hour for two hours, patient having had nothing to eat or drink for four hours previously. The quantities passed were as follows :—

> 1st quarter of an hour, 2¾ drachms.
> 2nd ,, ,, 2¾ ,,

He was then given fifteen minims of the one per cent. nitro-glycerine solution in a drachm of water.

> 3rd quarter of an hour, 12 drachms.
> 4th ,, ,, 16 ,,
> 5th ,, ,, 6¾ ,,
> 6th ,, ,, 8¾ ,,
> 7th ,, ,, 5¼ ,,
> 8th ,, ,, 3 ,,

The times were accurately taken, and in no instance was any of the urine lost. The increased secretion was obviously due to the drug. It is noteworthy that the maximum increase was not till the second quarter. Every specimen was examined as it was passed, and they were all free from sugar and albumen. The quantity was too small to admit of the specific gravity being taken by the urinometer, except in the case of the fourth quarter, when it was found to be 1003. It should be mentioned that this patient was very insusceptible to the action of the drug, and he experienced none of the ordinary symptoms from this dose.

In another observation on the same patient the results were still more striking. The same method of collecting the urine every quarter of an hour was adopted, and the following figures were obtained:—

			Sp. gr.	Pulse.
1st quarter of an hour, 4 dr.	....	—	....	64
2nd ,, ,, 10½ dr.	....	1003	....	64

Given twenty minims of one per cent. nitro-glycerine in one drachm of water.

			Sp. gr.	Pulse.
3rd quarter of an hour, 7 oz.	....	1000	....	80
4th ,, ,, 7½ oz.	....	1000	....	76
5th ,, ,, 1 oz.	....	1002	....	72
6th ,, ,, 7 dr.	....	—	....	68
7th ,, ,, 4½ dr.	....	—	....	64

The acidity of the urine varied inversely as the quantity passed. Thus, before the administration of the drug, it was distinctly acid, during the third and fourth quarters it was almost neutral, the acidity then gradually returned, till, in the seventh quarter, it was as marked as it had been at first. No sugar or albumen was detected either before or after the administration of the drug. The figures given under the head of pulse are averages of several observations made during each quarter of an hour. No subjective symptoms of any kind were produced. The experiment was commenced at ten in the morning, and patient had had nothing to eat or drink since breakfast at six. This epispadiac man was curiously insusceptible to the action of the drug as far as subjective symptoms were concerned. I gave him the one per cent. nitro-glycerine solution on ten different occasions, in doses of 3, 4, 4, 6, 12, 15, 15, 20, and 25 minims, without causing him a moment's pain or uneasiness. He never complained of headache, or beating or throbbing in any way, and yet the influence, both on the pulse and on the secretion of the urine, was well marked. Even the small doses affected the rate of his pulse. Thus, on one occasion, his pulse was taken every minute for eleven minutes, the average being 68. He was then given a little water in a medicine glass—a practice always followed in these observations—to test the effects of expectation. The pulse remained constant at 68 during the next five minutes, and six minims of the one per cent. solution were then given in water. In a minute and a half the pulse had risen to 76, and this increased rate was maintained for the next fifteen minutes, when it sank again to normal. On another occasion his pulse, taken on ten consecutive minutes, was found to be 80. He was then given twenty minims of the one per cent. solution in water. Half a minute after the pulse was still 80, in one and a half minute after it was 96, and in two and a half minutes after it was 100, the average of the eight minutes following the administration of the drug being 96.

Such were the results of the ten series of observations on this man—negative as regards his own sensations. As a final experiment it was decided that he should take a larger dose. At 11.51 A.M., sitting still in the cool laboratory, and having had nothing since an early breakfast, his pulse was 76. At 11.55 min. 30 sec. he took half a drachm of the one per cent. solution in a little water. At 11.56, pulse 76; at 11.57, 92; at 11.58, 96; soft and regular. At 12.4 he commenced yawning violently, and said he felt very sleepy. At 12.7 the pulse fell to 68, the yawning ceased, and he became very pale and complained of nausea. He was found to be perspiring freely all over the body, and was so hot that he kicked off his boots. The nausea lasted till 12.10, when the colour had returned to his face, and he said he felt all right again; pulse 76 to 80. There was no headache, and even a sharp run upstairs failed to produce any feeling of pulsation.

During the last four years I have employed nitro-glycerine in the treatment of many cases of angina pectoris, and with marked success. It is not necessary to reproduce them all in detail, and I have consequently selected a certain number which will serve as examples of the points I wish to illustrate.

The following was the first case of angina pectoris treated with nitro-glycerine :—

·CASE I.—William A., aged sixty-four, first came

under observation in December 1877, complaining of intense pain in the chest, excited by the slightest exertion. It was distinctly paroxysmal, patient being perfectly well in the intervals. The first attack was experienced in September 1876. Patient was at the time in his usual health, and was, in fact, out for a day's pleasure in the country. The pain seized him quite suddenly when walking. It was a most severe attack —as severe a one as ever he experienced in his life. It caused both him and his friends great alarm, and they were most anxious that he should return home at once. He cannot tell at all what brought it on; he had been enjoying himself very quietly; it was not by any means a cold day, and he had not been running, or even walking faster than usual. He remained perfectly well until the following April, when he experienced another similar attack; and since then he has been suffering from them with increasing frequency. From September 1877, they had been a source of constant anxiety, and it was only by a determined effort that he could continue to follow his occupation.

The attacks usually commence with a feeling of warmth, then of heat, and then of burning heat, in the chest, immediately followed by a heavy pressure, from the midst of which proceeds an acute pain, so that in a moment the whole chest seems as if it were one mass of pain. It is almost impossible, he says, to describe it, for he never felt anything like it before. The pain is first experienced at a small spot on either side of the sternum, corresponding to its junction with the fourth costal cartilages. From the chest the pain flies to the inner side of the arm, at a point midway between the shoulder and the elbow. It runs down as far as the elbow, but never to the fingers. It is not more severe on one side than the other. During the seizure, the patient suffers most acutely, and feels convinced that some day he will die in an attack. He usually experiences some shortness of breath at the time, but there is no feeling of constriction about the chest. He can speak during the seizure, though with some difficulty. The attacks are not accompanied by any sensation of warmth or chilliness, but patient is under the impression that he grows pale at the time. These attacks are induced only by exertion in some form or other, most commonly by walking, and especially by walking fast. Walking up hill is sure to bring on a seizure. Stooping down has a similar effect, and the act of pulling on the boots will excite a paroxysm almost to a certainty. He is almost afraid to stoop down, and when he wants to pick up anything from the floor, he goes down on his hands and knees. He has a slight cough, but although it shakes him at times it never brings on the pain. The attacks are not excited by food, but exercise taken after meals is more likely to induce them than when taken on an empty stomach. Patient has noticed that they are far more readily excited immediately after breakfast than at any other period of the day. They are more readily induced, too, after an indigestible meal than at other times, but patient is quite clear that no amount of indigestible food *per se* will bring on an attack. The paroxysms, as a rule, last only three or four minutes,

but occasionally from twenty minutes to half an hour.
If they come on whilst he is walking they always continue till he stops. Patient finds that stimulants afford
no relief. In the intervals between the attacks, patient
is perfectly well, and he feels that if he could only remain absolutely quiet the whole day long he would be
quite free from pain. Practically, as he is obliged
to be out and about, he has several attacks, on an
average six or eight every day. At the time of coming
under observation, the seizures were rapidly increasing
both in frequency and severity. His family history
was fairly good. His father died at the age of eighty-
three, and hardly had a day's illness in his life. His
mother died of phthisis, but only, patient says, through
catching cold, hers not being a consumptive family.
He lost two brothers—one at the age of eighteen, from
consumption, and the other in the tropics, cause un-
known. He has two brothers and one sister living,
all well. There is no family history of gout, asthma,
fits, heart disease, or sudden death. He has four
children, one of whom (a boy) is consumptive, and
another (a girl) subject to facial neuralgia. Patient
is a bailiff by occupation, and is a remarkably intelli-
gent man. He is a cool, clear-headed fellow, but little
prone to talk of his sufferings, although they are at
times very severe. He has travelled much, and has
lived in Egypt, Turkey, Italy, and Greece. For the
last thirty years he has been accustomed to lead an
active out-door life, seldom walking less than fifteen
miles a day, often very fast. He has, he says, done a
great deal of hard work in the way of pleasure. He

usually smokes about two ounces of bird's-eye in the
week, and has done so for years.   His health has
always been remarkably good, and with the exception
of rheumatism ten or twelve years ago, and pleurisy
seven years ago, he has never known what it is to be
laid up.   He has never suffered from gout.   On a
physical examination, it is noticed that there is some
fibroid degeneration of the arteries, and there is slight
hypertrophy of the left ventricle.   There are no signs
of valvular disease, and there is nothing to indicate
the existence of aneurism.   The urine was free from
albumen.

There could be no possibility of doubt respecting the
diagnosis.   It was a typical uncomplicated case of angina pectoris.

Patient was placed for a week on infusion of quassia,
in order that he might be observed, and also to elimi-
nate the effects of expectation.   It need hardly be said
that he derived no benefit from this treatment.   He
was then ordered drop-doses of the one per cent. nitro-
glycerine solution, in half an ounce of water three times
a-day.   At the expiration of a week he reported that
there had been a very great improvement.   The attacks
had been considerably reduced in frequency, and, for
two or three days he had had only one attack—in the
morning after breakfast.   The attacks, when they did
occur, were much less severe.   He found, too, that a
dose of medicine taken during an attack would cut it
short.   He had tried it several times, and it had always
.   succeeded.   It would not act instantly, but still very
quickly; so that the attacks were considerably short-

ened. He was thoroughly convinced that the medicine had done him good, and said he was better than he had been since first he had the attacks. It was found that the nitro-glycerine, even in this small dose, had pro-duced its physiological action. Patient complained that for two or three days he had experienced a strange ful-ness in his head, with a sense of pulsation. The pul-sation was felt chiefly in the temples, but also across the forehead. It caused him no positive inconvenience, and he evidently had no suspicion that it was due to the medicine. The dose was then increased to three minims, and patient found that this gave him more speedy relief. On two days during the week he had no seizure at all—a most unusual circumstance. Patient had adopted the plan of carrying his medicine with him in a phial, and taking a dose if an attack seized him in the street. It never failed to afford him relief. The beating had increased considerably in in-tensity, and was described as being a "kind of a pulse." Patient had discovered the fact that it was produced by the medicine. It came on immediately after each dose, and lasted about a quarter of an hour. It was now experienced chiefly across the forehead. Patient continued steadily to improve, and the dose was gradually and cautiously increased. With the increase in dose the pulsation became more severe, lasting from twenty minutes to half an hour. When twelve minims were given every three hours it became a positive inconvenience:

On January 14th, the dose of the nitro-glycerine solution was increased to fifteen minims every three hours. A few days later, he had a "kind of fit" immediately after having his medicine. The pulsation came on as usual, but was quickly followed by headache and pain at the back of the neck. His speech "began to go off," and he felt that he would have lost his senses had they not given him tea and brandy.

Patient took the fifteen-minim dose every three hours from the 14th to the 28th of January, but on the latter date had two "bad shocks." He took a dose of medicine in the morning as usual, and felt the customary pulsation, which passed off after about half an hour. An hour and a half later, he experienced a sensation as if he would lose his senses. He did not fall, but had to catch hold of something to prevent himself from so doing. It did not last more than half a minute, and there was no pulsation with it. The other seizure occurred later in the day, and was of the same nature. Patient attributed these attacks to the medicine, and was in no way alarmed by them. He thought it advisable, however, to reduce the dose by a third, and henceforth had no return of the fits. At this time, his anginal attacks were so thoroughly kept in check by the nitro-glycerine, that they gave him comparatively little inconvenience. He always carried his bottle of medicine with him, and immediately on experiencing the slightest threatening of an attack, he took a sip. Relief was certain, for even when it did not at once cut short the attack, it eased the pain so considerably, that he was able to go on walking. For two months longer he continued the ten-minim dose, sometimes taking a little more, and sometimes less.

The attacks became gradually less severe and less frequent, the dose was reduced to one minim, and in April he was so much better, that he was able to do without the nitro-glycerine entirely. He continued to attend from time to time until the following September, taking nothing but cod-liver oil, and a little tonic, and during the whole of that time, he had not a single attack. In July, 1879, he was seen again, and said he was well and strong. He had had a good deal of trouble and anxiety, but his health was excellent. He had had no return of the attacks. He could walk six miles at a stretch at a good pace, and found no difficulty in going up hill. He had gained weight, and ailed nothing. I still hear from him occasionally, and he is now (January, 1882) well and strong, and has had no return of his old attacks. In this case, the cure has been complete, and it appears to be permanent.

CASE II.—The second case was that of Mrs. H. S., aged 53, who first came under observation in January, 1878. She is a married woman and the mother of eight children. She complained of a " strange sensation " in her chest, over her heart, coming on in fits several times a day. It was not a pain, she said, at least not an ordinary pain ; it was something more than that—it was "just as if the life were going out of her." The attacks would last only two or three minutes at a time, but she seemed as if she could not get her breath, and they frightened her. She could just say " Oh dear !" or something like that, but nothing more. She would usually put her hand over her heart and press hard, and that seemed to relieve her. She feels quite cold

D

during an attack, and her friends tell her she gets pale in the face. The sensation is referred to a spot corresponding in situation to the point of maximum intensity of the heart's beat. It always keeps in the same place, and never flies to the shoulders or runs down the arms. In the intervals of the seizure she is perfectly well. There is no flatulence, nausea, vomiting, numbness in the arms, or vertigo, and the attacks are not followed by any discharge of urine. Patient never has an attack when quiet. The slightest exertion will bring one on : going upstairs will always do so, and even if she goes up very slowly she is sure to get an attack. She does not often get them on level ground, unless walking fast, and then she gets them. Going up-hill brings them on much more readily than walking on level ground. She can always tell, she says, when the ground is rising ; she knows directly. Shaking up a bed will bring on the pain at once. She dare not do it now, that is a great bother to her. Any little exertion is enough, as, for example, putting on her jacket or reaching up to the clothes line. Stooping down to lift anything brings them on, but not simply stooping down, as in pulling on her boots. Leaning back is certain to bring them on ; the least excitement will do so—in fact, anything that worries or upsets her. They are not in any way influenced by food. Cold feet will not bring them on, nor will a hot room. These attacks commenced at the beginning of last summer (1877), but were not so bad as they are now. They worried her a good deal, lasted on and off for two or three months, and then went away. She cannot tell at all what brought them on.

They returned on the following November, and have been getting worse ever since. Now she usually has seven or eight attacks a day, but the number depends very much on what she has to do. For some time past they have been gradually increasing in frequency, and are now far more readily excited than formerly. Her general health is fairly good. She has had a bad cough every winter for the last eighteen years. What with the cough and the children, she has never been very strong. She has never suffered from gout or anything like it. Patient's father died of gout and bronchitis. He had suffered from gout since he was twenty-one, and had large chalk stones. He was addicted to drink all his life more or less. His father and brother died of asthma. Patient's mother died in confinement, and she has no brothers or sisters. She lost one of her children from bronchitis and another from consumption. None of them ever had fits or St. Vitus's dance. On a physical examination, marked arterial degeneration is noticed. There is slight emphysema. There are no signs of aneurism and none of valvular mischief. Urine normal.

Here, again, little doubt was entertained respecting the diagnosis. It was not a typical case of angina pectoris perhaps, but it assimilated more closely to that type of disease than to any other. There could be no doubt about the reality of the patient's sufferings.

After a preliminary course of camphor-water, the patient commenced taking nitro-glycerine on February 4th. She was ordered one drop of the one per cent. solution in half an ounce of water every four hours.

In three days she reported that the pains had occurred less frequently; that they did not last so long. The pains were much shorter, and " there was a good bit of difference." She complained that the medicine had given her " such a strange sensation." It gave her " a kind of pain inside her head," and brought on a throbbing across her forehead just where the hair begins. After each dose she felt powerless for about ten minutes and had to sit down, feeling that she could not do anything. The dose was then increased to four minims every four hours, and this gave very marked relief to the anginal symptoms. The pains, she said, were very much better, and a dose of the medicine would always cut them short, almost at once; they were less frequent, less severe, and did not last so long. She was no longer afraid to hurry about the house, and was able to perform many little household duties that had been long neglected. She spoke very positively as to the good the medicine was doing her, but at the same time complained that it affected her most powerfully. The throbbing in her head after the dose was very strong, and lasted nearly twenty minutes; it was accompanied by a darting pain, and she felt cold all over; she had to sit down, and could do nothing as long as it lasted.

The patient continued to improve, and on February 21st she said she had taken a long walk the day before, not only without difficulty but with pleasure. Under ordinary circumstances the exertion would have brought on an attack and she would probably have had to return home. The attacks are now experienced only once or twice a day, in spite of her getting about much more;

and they are very much slighter than formerly, not lasting half the time. She does not take much notice of them now and no longer has to stop and put her hand over her heart. Some days she is entirely free from them.

Curiously enough, although the dose of the nitro-glycerine had been gradually increased to ten minims every four hours, the patient complained less of the throbbing in the head. During the following week the dose was increased, first to fifteen and then to twenty minims every four hours. The effect of the larger dose was very marked. She said the medicine made her "feel very bad;" she was afraid of it, for she felt it to her very fingers' ends. She throbbed all over—fingers, toes, and all. It affected her powerfully, and she had to sit down on the bed for nearly three-quarters of an hour after each dose. It caused noises in her ears just like the rushing of water, and made her feel cold all over. Sometimes it produces curious fits of gaping ; she went on yawning and yawning, and seemed as if she would never stop. It never made her feel faint, and when it was over she felt quite well again.

The dose of the medicine was now gradually decreased, and on March 7th it was abandoned in favour of general tonics. The patient was seen again early in 1880. She had then a bad cold on the chest but had not had a single bad attack of pain. I heard of her only a few weeks ago (January 1882), and she was then well and strong.

CASE III.—R. A,, aged 61, a painter's labourer, was first seen on April 11th, 1878. Complains of a

pain in the chest, which comes on when he walks.
The pain is referred to the midsternal region, and is
said to cover an area about the size of a teacup. It
is a dull, heavy, tight pain. It begins in the chest,
and then passes through to between the shoulders.
During severe attacks it sometimes runs down the
left arm as far as the elbow; it never extends to the
lower extremities. It is excited by exertion, and chiefly
by walking. It comes on suddenly, and he is obliged
to stop and wait till it goes off. He may have to stop
for a minute or two, or even longer. It often returns
when he starts again. When walking it may come on
several times in the course of half an hour, until at last
it brings him to a full stop. If he walks fast it will
bring it on, and so will going up hill. His ordinary
work does not excite it, nor does stooping. He gets it
chiefly morning and night, going to and returning from
work. Has not noticed that it is more readily induced
after meals, and does not think that food influences it
in any way. When pain comes on he gets pale—so his
friends tell him. Does not feel anxious, and the attacks do not frighten him at all. They are not accompanied by palpitation, but, during the attack, "he feels
very full," "as if he must burst," or "as if his chest
wanted moving." Patient has "knocked about a bit in
his time," but has been "fairly steady." First he was
on a farm, then in the police, then a wheelwright, and
now he is a painter's labourer. When in the police he
was advised to resign on account of weakness of his
chest, but does not think his chest was really affected,
for he had no cough, and has always felt well and

strong. Is subject to gout, and had his first attack
about three years ago. No history of syphilis. Has
been a great smoker for the last forty years; used to
smoke an ounce or more nearly every night, especially
when on night duty, and it was always shag tobacco,
and the strongest he could get. He experienced his
first attack twelve years ago, when working on the
Thames Embankment. It was the same kind of pain
as he has now, but it went off in a week or two. A
year later he had a return of it, which lasted for a few
weeks. Eight years ago a fire broke out, and he ran a
mile and a half to fetch the engines. This brought on
the attacks again, and he has had them more or less
ever since. He has been getting worse during the last
year, and especially during the last few months. On a
physical examination, it was found that the pulse was
irregular both in force and rhythm. There was some
arterial degeneration, and a slight arcus senilis was
noticed. No organic disease of the heart or lungs
could be detected, and there were no signs of aneurism.
Patient had a peculiarly anxious look, which was very
noticeable. No albumen in the urine.

After a short course of camphor-water, patient was
ordered a drop of the one per cent. solution of nitro-
glycerine in half an ounce of water, to be taken every
four hours. Four days later the patient reported that
there had been a great improvement. The attacks
were much less frequent, and that morning he had
walked to his work without having a single seizure—
a thing he had not done before for he could not say
how long. The attacks at night going home were just

as frequent, and he did not think they were less severe when they did come on. He had never taken a dose of the medicine when the attack was on him, so he could not say if it would cut it short. After each dose of the medicine he gets a pain at the back of the head, which comes on in about ten minutes and lasts half an hour. Says it is almost the same kind of pain as he has in his chest—"a heavy dull pain;" no beating or throbbing; no pain across the forehead or at the top of the head. Sometimes gets a "choky sensation in the throat" after the medicine. A few days later patient came again, and stated that he was steadily improving. At this visit he was given a single dose of two drops of the one per cent. solution on a piece of sugar. It produced slight flushing of the face and a marked increase in the fulness of the arteries. The pulse, which had previously been 98, rapidly rose to 112. The flushing was in a few minutes followed by intense pallor, and patient complained of feeling faint. He had to be supported to the sofa, his pulse was found to be very feeble, and it was a quarter of an hour or more before he was sufficiently recovered to stand alone. The patient was directed to continue the one-drop dose every four hours, and to take an extra dose when he felt the pain coming on. A week later he said he thought he was nearly well. For four days he had not had a single attack, although he had had a great deal of walking to do. When he felt any indication of the onset of the pain, he took a sip of his medicine, and it was all gone in a moment. He could walk to his work without the slightest difficulty, and even coming home at night gave him no trouble. The other day he walked the best part of a mile in a shower of rain quite briskly, and was none the worse for it. After each dose he experiences a pain at the back of the head and also over the forehead. A week later the dose was increased to two minims every four hours, and this was taken without difficulty. The medicine, he said, did not upset him at all. It had done him a deal of good, and he did not know what he should do without it. The dose was gradually and cautiously increased to eight minims every four hours. This was taken without difficulty, patient remarking that it did not upset him as it used to do. He was quite free from the attacks as long as he continued taking the medicine, but they returned immediately he discontinued it. He still attends at long intervals to report himself, but is practically well.

I have to thank Dr. Ringer for his kindness in having frequently examined these and other patients.

The next case is of interest, from the fact of the patient himself being a medical man. The treatment was conducted entirely by correspondence.

CASE IV.—In February, 1879, Dr. J. wrote to me saying that he was suffering from angina pectoris and that his symptoms bore a close resemblance to those of W. A., (Case I). In May, 1878, being then 65 years of age, he was suddenly seized, whilst hurrying up hill to attend an urgent case, with a severe pain under the sternum, which gradually extended all over the chest and down the left arm, compelling him to stop and remain motionless for some minutes. He subsequently had several similar attacks at short intervals. At first they came on only at night, awaking him from sleep and lasting sometimes from twenty minutes to half an hour. The pain was usually accompanied by a sense of oppression as if the chest failed to expand properly, or as if sulphur had been inhaled. A little later he found that the attacks were always induced by walking, or by hurry or excitement of any kind. Going up hill would be sure to bring it on, whilst riding could not even be attempted. Dr. J. had been carefully examined by several medical men by whom he was assured that there was no organic disease of either heart or lungs. He was under the impression that he was gouty, although he had never had a developed attack. The nitro-glycerine was at once commenced, with what success will be gathered from the following extract from Dr. J's. letter to the *Lancet*, April 19th, 1879. He says :—" I began to use the medicine at first in small doses—two minims of the one per cent. solution of the nitro-glycerine, every three or four hours during the day. I always found relief if I took the dose when I felt the first threatening of an attack, and the paroxysm was staved off. I continued to take the two minum doses regularly every three or four hours for about four days; and, as the attacks did not trouble me so much, I began to diminish the frequency of the dose and only took it when I felt an attack threatened. I always carry an ounce and a half bottle of the diluted solution in the breast pocket of my coat ; the bottle carefully marked for six doses, each dose containing five minims of the one per cent. solution. If I feel an attack coming on I apply to my bottle and at once feel that I am saved a paroxysm. The action of the medicine seems to commence the moment it is swallowed. It produces always a feeling of fulness in the head, singing in the ears, a sensation of pulsation all over, especially in the head; severe at the root of the nose as if epistaxis were threatened. I do not suffer from headache and the congested feeling soon goes off. It is a great boon to have a remedy in which you can have perfect confidence that the painful attacks can be controlled. I have not had any severe attack of the disease since I got the solution and began to take the drug, six weeks ago. I am, however, still unable to use my own powers of locomotion. I cannot walk without bringing on an attack of angina, especially if I go up hill or up stairs: even on the level I am obliged to walk slowly : but with my little bottle of solution in my pocket I can get about, feeling perfectly armed with a remedy to control an attack, if it should come on." I have since had several communications from Dr. J. who tells me, (November, 1881), that he has now retired from the active duties of a large practice and is almost free from his old attacks.

So many cases of angina pectoris treated successfully with nitro-glycerine have of late been published, both in this country and abroad, that it seems hardly necessary to add to the number. Much, however, may be learned from a study of unsuccessful cases, and I give in detail every case in which I have failed.

The following case unfortunately terminated fatally:—

CASE V.—William H., a coachman, aged 36, came

as an out-patient to the Royal Hospital for Diseases of
the Chest on the 10th of May, 1880, complaining of
pain in the chest on the slightest exertion. The pain,
he says, comes on quite suddenly, and often lasts five
minutes at a time, gradually passing away. He feels
it in front, "right in the hollow," (indicating ensiform
cartilage), and it then seems to go right through to between the shoulders, (at about the level of the angle of
the scapula), and after that it runs down the inner side
of the arms—both arms—as low as the elbow, but
never to the fingers. It takes away all power from the
muscles of the arms when it comes on, so that he
cannot hold anything in his hands. When the pain
comes on, there is a throbbing in the chest, which
mounts up right into the throat. The chest feels as if
it were loaded. He is short of breath, too, and feels that
he must keep quite quiet. It does not make him feel
frightened in any way. He has been told that he gets
very pale when the pain is on him.

The attacks are always brought on by exertion. If
he keeps quite quiet he hardly ever gets an attack, but
the slightest hurry or excitement will do it. Stooping
down is always sure to bring it on, such as trying to
wash his face at the sink, or trying to clean his boots
on. Putting the collar on the horses will bring it on,
and even putting on his own great coat. If his car-
riage is ordered out in a hurry, he has to get one man
to pull it out for him, another to harness the horses,
and a third to help him on with his own great coat. If
he were to attempt to do anything himself, he would be
done up in a minute and fit for nothing. If he gets

a pair of pulling horses it tries him very much, and "starts his chest" at once, but with an ordinary quiet horse he can get along very well. Walking will often bring on the attacks; yesterday he had to go for a newspaper, it was only a five minutes walk, but it took him a quarter of an hour to get there. He cannot walk fast, and is obliged to take things very easily. Directly he attempts to bustle about it is all up with him. As for running, it is out of the question, if he were to try to run ever such a little he would get the pain at once, he could'nt do it to save his life. Excitement he thinks will bring it on, but since the commencement of the present illness, he has taken good care not to get excited in any way. Tea or coffee, or lemonade or soda-water, will not bring it on.

He has noticed that the pains are most likely to come on after he has had a meal, but if he can remain quiet after his meal he is all right, and is none the worse for it. The pain is more likely to come on after breakfast, on moving about, than after any other meal. He has never had an attack at night.

As a rule the attacks come on quite suddenly, but sometimes he can feel the pain coming on, and then he "stops and rests, to prevent it coming on too tight," he might stop for a minute or two and lean against anything, but when time is not an object, he stops for five or ten minutes, for he never feels quite comfortable and safe till then.

It is impossible for him to say how many attacks he has in the day—it depends entirely on what he does.

For some time past he has been feeling very queer,

but he had no bad attack till about four or five weeks ago, when putting on the harness began to bother him, and he has been getting worse ever since. Three weeks ago, he knocked off work on account of the pain, and rested for a fortnight, he was nearly well at the end of that time, but directly he went back, he felt it worse than ever. He is now at work, but has very great difficulty in doing anything.

Patient has been a strong man all his life. He has worked for a horse-dealer, and has been accustomed to drive young untrained pulling horses. He never had rheumatic fever, and never gout. He has been a great smoker—about half an ounce a day, but the last four years he has been in a place where he could not smoke regularly and has not had half an ounce a week. He has lost flesh a little. On physical examination of the chest, nothing wrong was detected. There was no murmur, and no sign of aneurism.

It was clearly a simple case of angina pectoris. To begin with, he was given as a placebo, a mixture of sulphate of magnesia and peppermint water, and the attacks continued as usual. On the 13th of May he was ordered the following :—

Sol. Nitro-Glycerine 1 per cent. ℔ 80.
Aq. ad. Oj, M.

A teaspoonful gradually increased to a table-spoonful every three hours, with an extra dose at the commencement of each attack.

At the expiration of a week he reported that he had taken all the medicine. He said that even a teaspoonful caused a curious throbbing in his head, and that

when he took a table-spoonful it was worse. The
medicine would not stop the pain if he kept on walking, but if he stopped it would send it away quickly.
The dose was then increased to two minims of the
solution, and from that he derived much more benefit.
On the following day he walked two miles without an
attack, whereas usually he could not walk a quarter of
a mile without having to pull up. He felt so much
better, that one day he walked quite fast; the pain
seized him, he stopped and took a dose of medicine,
and was soon all right again. On the 24th, the dose
was increased to six minims, and on the 27th, to ten
minims. On the larger dose he improved very rapidly.
He walked two miles and a half, and had to stop only
once just for a little rest. He thought the pain was
coming on, so took a good pull at his medicine, and
was soon all right again. He can put on his boots
himself, and feels so much better that he thinks he
could go back to work again. He says the pain never
goes to the shoulders now, and the attacks are much
less severe than they used to be. The medicine causes
a good deal of throbbing in his head, but he can put up
with that. He takes on an average ten doses a day,
equal to a grain of the pure nitro-glycerine. On the
31st, the dose was increased to fifteen minims in the
half ounce of water, and of this he took a pint in four
days. He several times walked from two to three miles
straight off, at a good pace, and felt none the worse for
it. He complained that the medicine affected his head
more "it got right up to the top and kept on throbbing
there." For the next eight days the attacks were less

severe and less frequent, and he took only thirty-two ounces of the mixture. He then said he feared the medicine was losing its effect, for the attacks came on on the slightest exertion. On the previous day he had a bad attack, from straining, and could not move for ten minutes. He found too that suddenly turning over in bed would bring on the pain. The dose was then increased to twenty minims every three hours, and this he said did him much more good. The stronger it was the more quickly it checked the pain. For four days he had not a single attack, and felt so much better and stronger in every way, that he determined to go back to work. On the following day, June 15th, he was driving a drag, when he stopped and appeared to be in pain. He was taken into a house, and managed with difficulty to give his name and address, and ask that his wife might be sent for. He then became unconscious, and in a few minutes was dead.

He would in all probability have lived much longer had he not determined to resume work.

CASE VI.—J. H., 48, a boatman, was first seen Oct. 6th, 1880. He complained of a most acute pain in the chest, coming on in fits, starting from a point which he indicates—about half way between ensiform cartilage and umbilicus, and then passing upwards to the shoulders, but not running down the arms. It lasts usually ten minutes, but sometimes twenty, and takes away all power from him, so that he has to sit down. He is employed on the canal, but can no longer work as the slightest exertion brings on the pain. The number of attacks in the day depends entirely on what he does, or rather tries to do, but he has had as many as twenty in the twenty-four hours. He cannot walk a mile without having to stop five or six times. He has to stop absolutely, for he feels timid and faint, and sweats all over. His first attack was seven weeks ago, and up to that time he had been a perfectly healthy man. He is not a smoker. A careful examination of the chest failed to detect any signs of disease. The nature of his illness was explained to him, and he consented to give up work entirely, although he said it was absolutely necessary that he should go the present trip as it was too late to provide a substitute. He was ordered drop doses of the one per cent. solution, with spirits of chloroform and peppermint water, the dose to be increased to four drops if necessary. Nine days later, he called on his employers to say that the medicine had done him a great deal of good, that he was better in every way, and that he was just off to see me, to get a further supply. As he passed the window he dropped down dead. .

The following case also terminated fatally, but it was not a simple case of angina pectoris, there being extensive cardiac mischief.

CASE VII.—James O., a labourer, aged 38, came under my care as an out-patient at the Royal Hospital for diseases of the Chest, in January, 1880, complaining of pain in the chest. His story, told very much in his own words, is as follows:—He has an attack almost every time he goes out. For the first hour when walking he gets it every three or four hundred yards, so that it is difficult for him to get along, and he has to

E

go very slowly. It is a severe pain in the chest, on the left side, over the heart, it begins there and then goes into the left arm and runs down to the wrist. When it seizes him, it lasts for a minute and a half or more, and he has to stand quite still, or he pretends to be looking in the shop windows, so that people may not notice him. Walking or exercise of any kind brings it on, so that he has done no work for the last three weeks. Going up hill is very bad for him, he is sure to get it, and cannot walk so far going up hill as he can on level ground. Stooping down does not bring on the pain. If he ventures out directly after a meal he is more likely to get it than if he waits for a time, and it is always more likely to come on when first he goes out. He often gets the pain when undressing, but never when actually in bed. He finds that the best thing to ease it is something hot—hot coffee or whisky and water, for example. Cold things always make it worse.

His first attack was about two months ago, he was carrying a board up a high ladder and before he got to the top the pain seized him in the chest on the left side, palpitation came on and he had a hard job to get to the top. He got his mate to help him, and when he got to the top and was safe he stood there for close on five minutes trying to get right again and not able to move a step. It passed off after a bit and he kept on with the work that day, for it was a light job he was on. The second attack was five weeks ago, he was carrying a shift for the plasterers upstairs, when the pain came on so severely that he had to knock off at

once, and has not done a day's work since. On being
questioned about his mode of life, he said that for the
past three months he had been pretty steady, but
before that he used perhaps to take a drop more than
was good for him. His way he explained was this: he
used to take a good lot all at once and then go without
for a week or two. He drank beer mostly, a gallon or
more some days, but very little spirits, except perhaps
" a go of whisky now and again." He had been, he
said, a pretty good smoker in his time, he was twenty-
three before he began and then he took half an ounce
of shag a day, and perhaps chewed besides. He was
always healthy before this and was never laid up in
any way. No gout or rheumatism, and no rheumatic
fever. No family predisposition.

The patient had a curiously anxious care-worn appearance, and always spoke slowly and quietly, so as
to avoid the slightest hurry or excitement. He says
he feels frightened if anyone speaks to him quickly or
sharply. On examination of the chest, a loud double
aortic murmur was detected, but there was no increased
area of dulness. Urine, sp. gr. 1024, slight trace of
albumen. No elevation of temperature.

He was given a mixture containing spirits of chloroform, spirits of ether, and compound tincture of laven-
der, but it failed to ease the pain.

On January 8th, he was ordered two minims of the
one per cent. alcoholic solution of nitro-glycerine, in
half an ounce of water, to be taken every three hours,
with an extra dose at the onset of each attack of pain,
three days later he reported that the medicine eased

him at once—directly he had taken it. He had just
as many attacks but could stop them by taking the
medicine. He had not experienced any inconvenience
from it, and the dose was increased first to four minims
and then three days later to eight minims. The larger
doses did better than the small. If the pain came on
in-doors, and he took a dose, it was arrested at once,
but if he were out-doors, and tried to go on walking,
the pain soon came back in spite of the medicine. On
the 19th, the dose was increased to 14 minims, every
three hours, and that very night he had two most
severe attacks whilst in bed—the worst he had ever
had. They came on when he was asleep, the first
lasting he thinks half an hour, and the second a quarter of an hour. He took a dose of the medicine both
times, but it gave him no ease. For the next day or
two he had fewer attacks when out and about, and
could walk farther without getting the pain. He complained that the medicine was very cold, and said that
cold things were very apt to bring on an attack. He
always kept his bottle close to him so that it might get
warm before he took his dose. He was next ordered
$\mathfrak{m}$xx in half an ounce of cinnamon water every three
hours. The attacks became more and more frequent
so that he was hardly ever free from them. One evening when out walking he had an attack that lasted a
good half hour, he managed to walk on slowly, but had
to stop seven or eight times in half a mile. When he
stopped the pain was a little easier, but directly he
attempted to move it came on again. He took two
doses of the medicine with about a quarter of an hour's

interval, but it gave him no relief at all. It will stop
the attacks he has when in-doors, but not the out-door
attacks sufficiently to enable him to go on walking.
He still experienced no difficulty in taking the medicine. On the 26th the dose was increased to half a
drachm of the one per cent. solution every three hours.
The pains now came on not only on exertion but also
when he was perfectly quiet. One night he took five
doses from the time he went to bed to the time he got
up. It eased the pain, but he had to take it almost
continuously to get much relief. It was now determined to give nitrite of amyl a trial and he was given
a bottle and told to inhale freely at the onset of each
attack. This certainly did him good—it eased the
pain, although it would not stop it if he tried to go on
walking. It did him most good at first and then it
seemed to lose its effect and get stale as if it had lost
its strength. He was given a fresh bottle, but it was
no better and did not stop the pain at all. The attacks
became more severe and more frequent, and he could
not even walk a hundred yards without getting the
pain. He was not even safe in bed, and one night it
lasted three-quarters of an hour straight off. He could
not say which did him most good, the nitro-glycerine or
the nitrite of amyl, but neither seemed to help him
much. He was tired and worn out and his life was a
misery to him. Once more the nitro-glycerine was
tried, this time in 45-minim doses, eight times a day.
At first it seemed to have an almost magical effect, and
one day he walked two miles straight off without any
inconvenience. The medicine he said seemed to have

regained almost all its own power, and for a time he seemed almost well. A week later, however, the pains returned and the dose was increased to a drachm of the one per cent. solution eight times a day. He complained that it affected his head, but it was not much and he could bear it. In another week the dose was increased to seventy-five minims. This gave him much more headache, but it checked the pains most wonderfully. The dose was now increased, first to ninety minims, and then to a hundred minims eight times a day. This large dose—eight grains of pure nitro-glycerine in the twenty-four hours gave him some headache, but it soon passed off. It eased the pain very decidedly, and he felt that he could not do without it, he "craved for it," he said, it worked round his heart and down his left arm and then he was easy. At the expiration of a week the dose was increased to one hundred and ten minims. On the following evening he had a very bad attack, or rather a series of attacks, lasting in all an hour and a half. He had some hot port wine and then some nitro-glycerine on the top of it, but it made him worse if anything. Three days later he came again to the hospital and was admitted as an in-patient. He was kept strictly in bed and as far as possible free from excitement. From March 25th to April 11th, he was given nothing but camphor water, and practically all treatment was abandoned—at all events temporarily. On the 12th he was allowed to get up, but the pain was so great on even the slightest movement that the nitro-glycerine solution had to be given again, although this time only in five

minim doses. On the 15th, the dose was increased to
fifteen minims, and on the following day he was dis-
charged, having benefited decidedly by the enforced
rest. On the 19th he came again, as an out-patient,
and seemed so much better that the dose was reduced
to ten minims. On the 21st he had a violent paroxysm
of pain whilst in bed and died almost immediately.
No post-mortem examination was obtained. The wife
stated that he had unfortunately run short of medicine,
so that in this the final paroxysm she had none to give
him.

The great point of interest in connection with the
case is the wonderful insusceptibility of the patient to
the action of nitro-glycerine. The following table
shows the quantity of the one per cent. solution taken :

Date.	Dose.	When Taken.	Total.
Jan. 8 to Jan. 12	2 minims	8 times a day	64 minims
„ 12 „ „ 15	4 „	„	96 „
„ 15 „ „ 19	8 „	„	256 „
„ 19 „ „ 22	14 „	„	336 „
„ 22 „ „ 26	20 „	„	640 „
„ 26 „ Feb. 2	30 „	„	1680 „
Feb. 16 „ „ 23	45 „	„	2520 „
„ 23 „ March 1	60 „	„	3360 „
March 1 „ „ 8	75 „	„	4200 „
„ 8 „ „ 11	90 „	„	2160 „
„ 11 „ „ 18	100 „	„	5600 „
„ 18 „ „ 25	110 „	„	6160 „
April 12 „ April 15	5 „	4 „	60 „
„ 15 „ „ 19	4 „	25 „	400 „
			27,532 „

= 275 grains of pure nitro-glycerine.

The patient took doses equivalent to over a minim of

the pure drug eight times a day with hardly any inconvenience. It will be seen that from March 11th to
18th, he took daily 800 minims of the one per cent.
solution, or 56 grains of pure nitro-glycerine in the
week. From March 18th to the 25th, he took 6161
minims of the solution, equivalent to over a drachm
of the pure drug. I have given nitro-glycerine to hundreds of patients, and never before met with anyone
who could take anything like that quantity, in fact,
from 15 to 20 minims of the one per cent. solution
may in the majority of cases be regarded as a dangerous dose. That the solution was active was proved
by administering it to other patients, in all of whom
the usual symptoms were produced. I had at the same
time under observation an angina pectoris patient who
could not take more than two-thirds of a minim of this
same solution without suffering from throbbing headache and persistent sleeplessness.

Another point of interest is the progressive increase
in the severity of the attacks. At first the nitro-glycerine did a great deal of good, but subsequently it
proved of comparatively little value. The slight benefit
derived from the nitrite of amyl is also worthy of note.

The severity of the attacks is indicated not only by
their duration and frequency, but also by their onset at
night during sleep. It is probable that in this case no
treatment would have proved of much avail, although
both the nitro-glycerine and nitrite of amyl acted as
palliates. It is unfortunate that in the last and fatal
paroxysm he should have had no medicine.

The next case presents many points of interest.

That a man should suffer from anginal attacks for thirty years is, I think, almost unprecedented. His susceptibility to the action of nitro-glycerine affords a marked contrast to the last case. I have included it amongst the failures, but it should be noted that the patient frequently stated that the nitro-glycerine had done him more good than anything he had ever taken.

CASE VIII.—S. W., 56, formerly a warehouseman, now a clerk, but lives chiefly on an allowance. Is subject to sudden attacks of pain at the chest, which seem to be at his heart. It is a sudden pain or cramp, he cannot describe it in any other way, but it is most severe. It always comes in the chest but not always in the same place. It generally flies to the left shoulder, but never down the arm. The attack lasts from five minutes (a short one) to an hour (a long one). He had it once for five hours, but that was a succession of paroxysms. He volunteers the statement that he feels as if he would die in an attack, and that he is afraid to go out on that account. The attacks occur most frequently out of doors, sometimes the least exertion will bring them on. Stooping down will bring them on more quickly than anything. Can manage gentle walking fairly well, but the slightest attempt to hurry brings on the pain at once, and he has to stop. Lifting or straining will always bring on a bad attack. Excitement has the same effect, and he never gets angry or quarrels. Depressing emotions readily excite it, and although he is a church goer, he has to avoid eloquent preachers. He takes a great interest in politics, but would not dare go to a political meeting.

He is more likely to suffer from an attack after break-fast or supper, than at any other time, and he rarely goes out for two hours after breakfast. Things that produce flatulence do him harm, he is afraid to take beer, and finds that the best thing is a little weak spirit and water, but he is most obstemious in every way. Sunshine and heat will bring on the pain, and so will extreme cold.

He has suffered from these attacks for thirty years, and on this point he speaks decidedly. Twenty-six years ago he was under the care of Dr. Risdon Bennett when the Victoria Park Hospital was at Finsbury, and his complaint was then diagnosed as Angina Pectoris. He has also been under the care of Sir Thomas Watson, Dr. Peacock, and most of the better known physicians and surgeons of the day. His attacks when first they came on were much of the same character as they are now, but they were not so frequent and did not last so long. He was once for a year without an attack. During the last four years the attacks have occurred much more frequently, and have been of longer dura-tion. He has now on an average two or three or per-haps half a dozen slight attacks in the day, and then a bad one about once in two days. He never goes a day now without an attack of some kind. His chest has been carefully examined many times, but nothing wrong can be detected; pulse 86, regular.

He was ordered two minims of the one per cent. solution of nitro-glycerine every three hours. Three days later he returned, saying he could not take it, it made him so ill. He related his experience of the first dose; it had no sooner gone down than it seemed to rise from the chest into the head, there was a tremendous sense of fulness, and he felt as if his head would burst. It seemed to bring on the pain. It kept him awake, and he was excited and restless. He was afraid to take it during the paroxysm for fear it should make him worse. He expressed a strong opinion that it would not suit him, and said he would rather not take any more. He says he is very susceptible to all drugs, digitalis in even the smallest dose upsets him at once, opium always keeps him awake, and he has never been able to smoke.

The dose was reduced to a quarter of a minim every three hours. He took this without much difficulty, although it still caused some fulness in the head. He thought it shortened the duration of the attacks. After taking it for a week the dose was increased to a third of a minim, and it was given in peppermint water. He again complained that it was too strong, and that it kept him awake. He says it certainly eases the pain, the attacks do not last so long, and there is not that terrible sense of terror with them. The dose had to be reduced to a quarter of a minim, but a week later it was increased to a half. It eased his pain, but the fulness in the head and the sleeplessness were so distressing, that he could not take it. It was thought possible that belladonna might counteract the head symptoms, and he was given a third of a minim of the nitro-glycerine solution, with fifteen drops of tincture of belladonna, but the pulsation was no less, and he could not take it. The belladonna was discontinued and the nitro-glycerine was gradually increased to two-thirds. A single dose of this would keep him awake for hours, and he says that night after night he never closed his eyes till four in the morning. He was given twenty-five grains of bromide of potassium, and five grains of bromide of ammonium, as a sleeping draught, but it failed to remove the sleeplessness induced by the nitro-glycerine. Hyoscyamus at bed-time was then given with the view of inducing sleep, but it did no good. At last he decided to give up the nitro-glycerine, but he had no sooner abandoned it, than the pains increased in frequency and severity, and he begged to have it again, saying that he could not do without it. On the whole he was satisfied that the treatment had done him good, and that he had derived more benefit from the nitro-glycerine than from anything he had ever taken. Large doses—by that he meant two-thirds of a minim of the one per cent. solution—made the pain worse. They brought on the headache, and that was followed by the anginal pain. Smaller or moderate doses suit him admirably—of that he is confident—and do him more good than anything.

The patient was now given Parke Davis & Co's. gr. $\frac{1}{100}$ nitro-glycerine pills. There was no doubt about their activity. He said they had not been down a minute before he felt the effect. He took five a day for a short time and derived considerable benefit from them. He was next tried with a mixture containing half a minim of the one per cent. solution and six minims of tincture of hyoscyamus in peppermint water, and this he took for a month also with advantage, the hyoscyamus apparently controlling to some extent the head symptoms, and preventing restlessness at night. The attacks were less severe and much less frequent. It was next determined to try nitrite of amyl instead of nitro-glycerine. He was given a half ounce bottle and taught to inhale it freely several times a day, especially at the onset of the attack. He tried it for a fortnight, but found that it did not ease the pain so effectually as the nitro-glycerine. He was then given a mixture of half a drachm of nitrite of amyl in two drachms of rectified spirit, three drops or more to be taken on sugar every three hours, and during the paroxysm. This too he tried for a fortnight, but the attacks became more frequent. Ergot was tried next, and without benefit. A mixture containing bicarbonate of potash, aromatic spirits of ammonia, spirits of chloroform, spirits of ether, and tincture of capsicum, also proved ineffectual. A course of tincture of cactus grandiflorus did him no good, and the same may be said of the sulpho-carbolates, and of Jamaica Dogwood. Finally, at his own request, he was again put on the nitro-glycerine, which undoubtedly did him more good than anything. He remained under treatment for exactly a year, and then ceased to attend.

The following case is of interest as showing that the existence of extensive cardiac mischief is no bar to the administration of nitro-glycerine in cases presenting anginal symptoms.

CASE IX.—W. H. W., aged 33, a printer came to the Hospital, on Nov, 25th. 1878, complaining of severe pain in the chest on exertion. His first attack was three months ago, early one morning as he was going home from work. He was in St. Paul's Churchyard at the time and had just come up Ludgate Hill. It was a severe pain in the chest, and lasted about three minutes. Two days later he had another similar attack and he has had them, on and off, two or three times a day ever since. They come on more readily than they did and have of late increased greatly in severity. He has been under a doctor, but as received no benefit.

The pain usually seizes him after exertion, the slightest effort will bring it on. An ordinary attack lasts only about two minutes, but a bad attack will last a quarter of an hour or more. Walking is sure to bring it on, especially walking up hill. He has to walk two miles and a half to his work and may have to stop sixteen or seventeen times, sometimes not more than eight times, but he never manages to walk all the way without stopping. Stooping down will bring on the pain, as for example, pulling on his boots, or washing his face. His work does not bring it on much, but then it is not heavy work. He finds, that he must not get excited or talk much or the pain comes on, and soon puts a stop to his chattering. Once or twice the attacks have come on during sleep awaking him in great agony. The pain is a sore raw pain, a distressing kind of pain as if something rose out of its place. He has to stand still when the pain comes, but does not have to hold on to things. He is usually pale during the paroxysm. The pain is limited to the chest, and does not extend to the shoulders or run down the arms. There is no cough, and no shortness of breath, and he has not lost flesh. He has been married 14 years and he has had two children. He had syphilis about 12 months before he married. He is a smoker and used to smoke an ounce and a half in two days, but of late he has reduced the quantity to about a fourth. He has chewed for the last 14 or 15 years, generally about a quarter of an ounce a day.* He has drunk a great deal, chiefly beer but also spirits. No gout, rheumatism or rheumatic fever.

There is an aortic diastolic murmur heard most distinctly at about the middle of the sternum. His chest is a little higher pitched on the right than on the left, but not much, and no importance is attached to it. No bulging; no thrill. There is a characteristic aortic pulse. No arcus senilis. Urine, sp. gr. 1020, no albumen.

At his first visit he was ordered five grains of Iodide of Potassium three times a day, but from this he derived no benefit. On December 2nd he was ordered two minims of the one per cent. of nitro-glycerine every three hours. He reported a week later that it stopped the the pain at once, and without causing him any inconvenience with the exception of a little headache. As an example of the benefit he derived from it, he mentioned, that on his way to the hospital, he had to stop only once, whilst a week ago he had to stop more than a dozen times. On December 16th the dose was increased to five minims every three hours and this he said suited him even better, and stopped the attacks instantly. They were not nearly so severe, and the pain

---

* This has been a source of great difficulty all through the case. He chews cut cavendish at his work and cannot give it up. He finds that coca leaves are the best substitute.

was gone in a moment, as soon as he had taken the
medicine. Two days when he was short of it he was
not so well. He progressed favourably having but few
attacks, until Jan. 20th, 1879, when he had three or
four very bad days. He had one bad attack which came
on at 5 o'clock in the morning and lasted half an hour.
The medicine although generally so useful, did him no
good in this particular attack. The dose was then increased to ten minims to be taken every two hours or
oftener if necessary. On this he improved rapidly, the
attacks being kept thoroughly in check. At first he took
about ℥ xiv of the one per cent. solution in the week,
but as he became accustomed to it he took it more frequently and his allowance was increased to ℥ xxi. In
April it was determined to try nitrite of amyl for a time,
experimentally. He was made to inhale it freely until
he was distinctly flushed and he was given a bottle with
directions to use it as an inhalation on the onset of an
attack. After a week's trial he said he did not get on
with it so well as with his old medicine. It flushed
him and relieved his attacks completely, but although
it acted even more quickly, its effects were transitory.
At his own request he returned to nitro-glycerine which
was again given in ten minim doses. His usual mixture was ℥ iii of the one per cent. solution to a pint of
water, the dose being a tea-spoonful. Of this he takes
on an average six ounces in the 24 hours sometimes a
little more and sometimes less. This is equivalent to
48 doses a day, i.e. 480 minims of the one per cent.
solution or 48 grains of pure nitro-glycerine. He was
allowed to take it very much in his own way. As a
rule he took it only during the paroxysm, the dose

being gradually increased till relief was obtained. He always carried the mixture about with him in a sherry flask, and by its judicious use he was nearly always able to check the attacks. The mixture was taken almost without interruption from June 1879 to January 1882. The quantity taken was considerable, for example, in a period of 126 consecutive days, he took 604·8 grains of pure nitro-glycerine, giving an average 4·8 grains a day. This is a long case and it is unnecessary to give an account of the progress from day to day, but the following figures will be of interest. In successive periods of six weeks the patient took an average weekly dose of 4ozs. (=19·2 grains), 6ozs. (=28·8 grains), 6ozs. (=28·8 grains), 6ozs. (=28·8 grains), 6ozs. (= 28·8 grains), 7ozs. (=29·6 grains), 5ozs. (=24·0 grains,) 5ozs. (=24·0 grains), 3·5ozs. (=16·8 grains), 3 ozs. (=14·4 grains), 4ozs. (=19·2 grains), 6ozs. (=28·8 grains), 4·5ozs. (=21·4 grains), 4·5ozs. (=21·4 grains,) 5ozs. (=24 grains), 5·5ozs. (=26·4 grains), 6ozs. (= 28·8 grains), and 5ozs. (=24 grains). The ounces refer to the one per cent solution, whilst the grains give the equivalent in pure nitro-glycerine. During a period of 102 weeks the patient took 1767 grains of the pure drug or considerably over two pounds. The benefit was most marked, the attacks became fewer and fewer and less and less severe, so that with the help of the mixture he could work almost as well as he did before. He rarely found it necessary to work more than four days a week, his pay being good, but he could do much more. For example, once when they were unusually busy, he worked 82 hours in 6 days, and on another occasion 36

F

hours at a stretch. For several weeks he averaged 66 and 68 hours a week, and that he considered was as much as he cared to do when he was well. It may be said that for a period of over three years he has lived on nitro-glycerine, at all events, he has been kept alive by it. He has had no other treatment, with the exception of a little phosphorus, a fiftieth of a grain, three times a day, from time to time. The phosphorus in no way relieved the pain, but I was in hopes that it might have done him good, and he thought that it was useful in bringing up the phlegm, if he caught a cold. I have examined his chest from time to time, but have detected no alteration in the physical signs. His general condition is good and he has not lost flesh. He still remains under treatment. Dr. Ringer very kindly gave me the benefit of his opinion in this case.

I have met with other cases of cardiac disease accompanied by anginal attacks in which benefit has been derived from taking nitro-glycerine.

CASE X.—A short time ago a gentleman called on me and told me that he had taken nitro-glycerine, in large doses for over a year for what he supposed to be angina pectoris. He had not had medical advice at all but had treated himself, and the result he assured me had been most satisfactory, for he had improved in every way. He had come, not to consult me professionally, but because he thought I might be interested in his case and should be glad the treatment had done him good. He gave me a long account of his sufferings, but as his description hardly seemed to bear out his diagnosis of angina pectoris, I ventured to ask permission to examine his chest, when I found he had a loud double aortic murmur. I communicated with his wife and advised her, to see that he had proper medical attendance. I have never seen or heard of him since and do not know whether he still continues the nitro-glycerine.

Nitro-glycerine often succeeds well in what may be called pseudo-angina. I am not referring to the pains in the chest and elsewhere, sometimes experienced by young or hysterical women, but to those cases, met with chiefly in men, where some or perhaps nearly all the symptoms of angina pectoris are suffered from, but the attacks are not typical in character.

In the following case, of which an abstract of the notes is given, the administration of nitro-glycerine was attended with success.

CASE XI.—L. B., soap maker, aged 42. Complains of pain in the chest on the left side, constant, but increased by movement, very severe at times, and occasionally so acute as to make him cry out; seems as if it would take his breath away; sometimes occurs between the shoulders as well, and not unfrequently runs down the left arm as far as the elbow. If walking and the pain comes on, he has to stop, but only for a few seconds, and then goes on again. The pain is increased by stooping down, as in putting on his boots. Any movement even turning in bed, will bring on the acute pain; but still he his never entirely free from it. He has it more or less acutely on moving. He has the very greatest difficulty in doing his work. Has been abstemious all his life; a smoker, but not consuming

F 2

more than half an ounce of tobacco a week. Has had gout thirty times or more during the last twelve years. Has had winter cough for about the same time. Never had these pains until this year. Has been gradually losing flesh for some months past. Physical signs those of emphysema; heart normal; no albumen in the urine. The patient was ordered a gentian-and-soda mixture, and this he took for a fortnight without the slightest benefit. The medicine, he said, did him more harm than good, The local application of belladonna failed to afford relief. He was then given drop doses of the one per cent. nitro-glycerine solution in half an ounce of water four times a day. A week later he reported that he had felt relief on the first day, and had steadily improved ever since. He could stoop down without getting the old attacks, and could walk about almost as well as ever. He had not the slightest difficulty in taking the medicine. He remained under observation for some time longer, but there was no return of the pain.

Another somewhat undefined case of the anginal type, benefitted by nitro-glycerine is the following :—

CASE XII.—H. C., aged 32, a bookbinder, says he suffers from attacks of acute pain in the chest, which come on suddenly, last a few minutes, and then pass away again. The pain begins just under the nipple, on the left side, but does not go to the shoulder or down the arm. It is a feeling as if someone were squeezing the heart. It comes on chiefly when he is at work and especially when he has labourious work to do. He rarely gets it at other times, never for example, on

Sunday. Walking does not bring it on, but he dare not walk fast, and he never runs, if he hurries a little in crossing the street, to get out of the way of anything it brings on a little pain, but not usually a bad attack. He does not think he ever had a decided attack in bed, but he often awakes feeling that something has happened, as if the pain were just passing off. His first attack was about eight months ago, when he was doing laborious work in a constrained position. The pain was so acute that he fell down, and his heart seemed to stop. Two or three days later he had another bad attack, and had to give up work for a time. He has been a smoker all his life, usually half an ounce of shag or twist a day, but often six ounces or more a week. On examining the chest nothing wrong could be detected. He was given at first only peppermint-water and on this there was of course no improvement. A drop of the one per cent. solution of nitro-glycerine was then, without the patient's knowledge, added to each dose and on this he rapidly improved, complaining at the same time that it gave him a headache.

A consideration of these cases will enable us to arrive at certain conclusions respecting the dose and mode of administering nitro-glycerine.

What is the best way to give it? I generally use the one per cent. alcoholic solution. It is a colourless almost tasteless fluid, one minim being equal to gr. $\frac{1}{100}$ of pure nitro-glycerine. It may be given in a drachm of water, or in any quantity that may be found most convenient. In some cases I have given it with a few drops of spirits of chloroform in peppermint water, and this mode of administration is advantageous when any-thing cold excites or intensifies the paroxysm, the spirits of chloroform and peppermint water not only taking off the coldness but relieving the flatulence which is so commonly an accompaniment of the attack.

At one time I thought a one per cent. solution of nitro-glycerine in ether might be useful, the ether itself helping to allay spasm. I gave it a fair trial, but it was not a success ; the patient complained of the constant smell and taste of the ether, and preferred the tasteless alcoholic solution.

Mr. Martindale kindly made me some pills composed of nitro-glycerine dissolved in cocoa butter. They are perfectly active, but if swallowed in the ordinary way do not act so quickly as the spirituous solution. If chewed, they are not very nasty ; they act promptly enough. They may be obtained of any strength from gr. $\frac{1}{200}$ to gr. $\frac{1}{10}$. My patient, W. H. W. (Case IX.), fully appreciated them and took sixty a week of the one-fiftieths for several weeks, in addition to a very fair allowance of the solution.

The nitro-glycerine pills—or rather pilules—made by Messrs. Parke, Davis, and Co., of Detroit, Michigan, are perfectly active, and are taken without difficulty. They are sent out in small bottles containing ten dozen or more, and may be obtained either sugar-coated or gela-tine-coated, the latter being preferable. They are made of five different strengths: gr. $\frac{1}{100}$, gr. $\frac{1}{50}$, gr. $\frac{1}{33}$, gr. $\frac{1}{25}$, gr. $\frac{1}{10}$, and they keep well. On taking one of the gr. $\frac{1}{100}$, gelatine-coated pilules, I found it produced the characteristic symptoms—pulsation, headache, &c.

W. H. W. took, in five days and a half, a gross of the gr. $\frac{1}{25}$. He has tried all these pilules, and finds they are all active. He has taken so much nitro-glycerine that he is quite a connoiseur.

Martindale's nitro-glycerine tablets made with chocolate are perfectly active and afford an excellent method of administering the remedy. They are a great convenience to patients who are not confined to the house, and who do not like to be seen drinking out of a bottle in the street. They act promptly and will ward off an attack. A great advantage is that they can be obtained of any strength. A patient of mine took them for over a year with much benefit.

The dose. On this point, it is impossible to lay down any definite rule. The cases here recorded serve to illustrate this. For example, J. O. (Case VII.), could take over 100 minims of the one per cent. solution at a dose, whilst S.W. (Case VIII.), suffered severely from half a minim. As a rule one minim is a safe dose to begin with, but it would be better to be quite on the safe side, and to give only half a minim at first. I nearly lost a patient by giving him gr. $\frac{1}{50}$ for the first dose. Another patient—a young lady—suffered from intense headache after taking a sixth of a minim of the one per cent. solution. This was not imagination, for she did not know what she was taking, and probably had never heard of nitro-glycerine. Delicate young women are undoubtedly more susceptible to the action of the drug than other people. Quite recently I made some observations on fifty women, taken at random, and of various ages, and found that ten per cent. complained of the medicine after taking doses of one minim of the one-per-cent. solution; on increasing the dose to two minims, thirty-two per cent. complained, whilst more than half suffered from headache after taking three minims. Although it is a good plan to begin with a small dose, there is no reason why it should not be increased rapidly. If half a minim produces no effect, give a minim at once, and go on increasing the dose until the patient complains. It must be re-. membered that the object is not to give a certain dose, but to get the physiological effect of the drug. After a time tolerance is established, and the patient is able to take more than at first. An over-dose might produce alarming symptoms, as in the case of W. A., (Case I), but the effects are commonly transitory, and the patient recovers before assistance can be obtained.

The frequency of administration. I generally give a dose every three hours, with an extra dose immediately at the onset of an attack. As there is no telling when the pain may come on, the patient should always carry the pills or tablets, or some of the solution in his pocket. One patient always keeps his supply of solution in a sherry flask. It should be remembered that when the attacks are severe, it may be necessary to give dose after dose in quick succession until relief is obtained.

Accessory treatment. In nearly all the cases here recorded, nothing was given except nitro-glycerine, but certain accessory measures might be resorted to with advantage. Thus the patient would be dieted, especial care being taken to avoid anything likely to produce flatulence, which so often accompanies anginal attacks. I should attach much importance to discontinuing smoking, for it will be seen that many of the patients had been in the habit of smoking to excess, (Cases I., III., V., VII., IX., XII.). I have sometimes given arsenic or phosphorus in addition to the nitro-glycerine. They often improve the patient's general condition, but do not in any way mitigate the severity of the attacks. It will be seen that in two cases nitrite of amyl did no good, but it must be remembered that they were exceptionally severe cases, and no one doubts the enormous benefit obtained from the inhalation of nitrite of amyl in most cases of angina pectoris.

Safety of nitro-glycerine. It has sometimes been urged against the employment of nitro-glycerine as a remedial agent that it is a dangerous explosive. This objection does not apply to the form in which it is ordinarily dispensed. The one-per-cent. solution is perfectly safe, and may be used without fear; in fact, most chemists keep a five-per-cent. solution. Nitro-glycerine may be kept dissolved in fat for any length of time with perfect safety. Mr. Martindale gave me an ounce of the basis from which his pills are made, and I made numerous experiments with it without getting it to explode. It was carried first to Westminster outside an omnibus, and then to the City by Underground Railway. It was placed on an iron plate, and a heavy weight allowed to fall on it from a considerable height. It was wrapped in paper and hammered for sometime on an anvil. It was stirred up first with a red-hot wire, and then with a lighted match, and was finally got rid of by throwing it out of a second-floor window.

Nitro-glycerine is sometimes used in neuralgia, megraine, epilepsy, asthma, whooping-cough, Bright's disease, sea-sickness, and other complaints.

INDEX.

Just published, medium 32mo, *cloth,* 1s. 6d.

WHAT TO DO

IN

CASES OF POISONING

BY

WILLIAM MURRELL, M.D., M.R.C.P., M.R.C.S.

*Lecturer on Materia Medica and Therapeutics at Westminster Hospital; Senior
Assistant Physician, Royal Hospital for Diseases of the Chest.*

Notices of the Press.

" Dr. Murrell truly says, ' few things can be more painful than to
be called into a case of poisoning, and not to know what to do.' In
this little book, plain and straightforward directions are given for
the treatment of the commoner poisons. The tables were originally
drawn up by Dr. Murrell for his own guidance, and there can be no
doubt that they will be extremely useful to others. It is a little
handy pocket-book which contains a series of extremely practical and
accurate directions for the application of antidotes in cases of the
treatment of poisons from the now great variety of poisonous agents
which are apt to give rise to fatal symptoms. It is the handiest and
most complete of the kind that we have ever seen."—*Brit. Med.
Journal.*

"The author gives instructions very concisely as to the course to
be taken in cases of poisoning with any of the mineral and vegetable
poisons, drawing his information from more recent sources than
most of the established tables of antidotes."...*Chemist and Druggist.*

" The object of Dr. Murrell's tiny, but useful, volume is ' to give
straightforward directions for the treatment of the commoner poisons '—*i.e.*, for the treatment of cases of poisoning by them. This he
has done in a clear, concise manner, and in a form calculated to be of
real service to practitioners. The points of general management to
be adopted in each case are given in separate paragraphs, and directions are supplied for antidotal treatment which embody the latest
knowledge on the subject of the antagonism of drugs."—*Lancet.*

"As might be expected from Dr. Murrell's training and ante-
cedents, this is an excellent and reliable little compendium, which
fairly fulfils its object of giving plain, straightforward directions for
the treatment of cases of poisoning. It is clearly arranged both as
to matter and type, and contains, in a very small compass, a large
amount of valuable information. We strongly recommend every hospital house-surgeon and resident pupil to procure a copy and make
himself master of its contents, after which he will not feel at a loss
what to do in almost any imaginable case of poisoning that may
suddenly present itself for treatment."—*Dub. Med. Journal.*

LONDON : H. K. LEWIS, 136 GOWER STREET, W.C.

CATALOGUE OF WORKS

Published by

H. K. LEWIS, 136 GOWER STREET,

LONDON, W.C.

G. GRANVILLE BANTOCK, M.D., F.R.C.S., EDIN.

Surgeon to the Samaritan Free Hospital for Women and Children.

A PLEA FOR EARLY OVARIOTOMY. Demy 8vo, 2s.

FANCOURT BARNES, M.D., M.R.C.P.

Physician to the British Lying-in Hospital; Assistant Physician to the Royal Maternity Charity of London, &c-

A GERMAN-ENGLISH DICTIONARY OF WORDS AND TERMS USED IN MEDICINE AND ITS COGNATE SCIENCES. Square 12mo, Roxburgh binding, 9s.

ROBERTS BARTHOLOW, M.A., M.D., LL.D.

Professor of Materia Medica and Therapeutics, in the Jefferson Medical College of Philadelphia, etc.

I.

A TREATISE ON THE PRACTICE OF MEDICINE FOR THE USE OF STUDENTS AND PRACTITIONERS. With Illustrations, large 8vo, 21s.

II.

A PRACTICAL TREATISE ON MATERIA MEDICA AND THERAPEUTICS. Fourth Edition, Revised and Enlarged, 8vo, 16s.

GEO. M. BEARD, A.M., M.D.
Fellow of the New York Academy of Medicine.

AND

A. D. ROCKWELL, A.M., M.D.
Fellow of the New York Academy of Medicine.

A PRACTICAL TREATISE ON THE MEDICAL AND SURGICAL USES OF ELECTRICITY; including Localized and General Faradization; Localized and Central Galvanization; Electrolysis and Galvano-Cautery. Third Edition. With nearly 200 Illustrations, roy. 8vo, 28s.

DR. THEODOR BILLROTH.
Professor of Surgery in Vienna.

GENERAL SURGICAL PATHOLOGY AND THERA-PEUTICS. In Fifty-one Lectures. A Text-book for Students and Physicians. Translated from the Fourth German Edition with the special permission of the Author, and revised from the Eighth German Edition, by C. E. HACKLEY, A.M., M.D. Copiously illustrated, 8vo, 18s.

G. H. BRANDT, M.D.

ROYAT (LES BAINS) IN AUVERGNE, ITS MIN-ERAL WATERS AND CLIMATE. With Preface by DR. BURNEY YEO. Frontispiece and Map. Crown 8vo, 2s 6d.

GURDON BUCK, M.D.

CONTRIBUTIONS TO REPARATIVE SURGERY; shewing its application to the Treatment of Deformities, produced by Destructive Disease or Injury: Congenital Defects from Arrest or Excess of Development; and Cicatricial Contractions from Burns. Illustrated by numerous Engravings, large 8vo, 9s.

J. B. BUDGETT, L.R.C.P. EDIN.

THE HYGIENE OF SCHOOLS; Or, Education Physically and Mentally Considered. Crown 8vo, 2s.

FREEMAN J. BUMSTEAD, M.D., LL.D.

*Late Professor of Venereal Diseases at the College of Physicians and Surgeons,
New York.*

THE PATHOLOGY AND TREATMENT OF VENE-REAL DISEASES. Fourth Edition, revised, enlarged, and in great part re-written by the Author, and by ROBERT W. TAYLOR, A.M., M.D. With 138 woodcuts, 8vo, 25s.

ALFRED H. CARTER, M.D. LOND.

*Member of the Royal College of Physicians ; Physician to the Queen's Hospital,
Birmingham. &c.*

ELEMENTS OF PRACTICAL MEDICINE. Crown 8vo, 9s.

P. CAZEAUX.

Adjunct Professor in the Faculty of Medicine of Paris, &c.

A THEORETICAL AND PRACTICAL TREATISE ON MIDWIFERY ; including the Diseases of Pregnancy and Parturition. Revised and Annotated by S. TARNIER. Translated from the Seventh French Edition by W. R. BULLOCK, M.D. Royal 8vo, 175 Illustrations, 30s.

JOHN COCKLE, M.A., M.D.

Physician to the Royal Free Hospital.

ON INTRA-THORACIC CANCER. 8vo, 4s 6d.

W. H. CORFIELD, M.A., M.D. OXON.

Professor of Hygiene and Public Health in University College, London.

DWELLING HOUSES: their Sanitary Construction and Arrangements. With 16 pages of Illustrations, crown 8vo, 3s 6d.

J. THOMPSON DICKSON, M.A., M.B. CANTAB.
Late Lecturer on Mental Diseases at Guy's Hospital.

THE SCIENCE AND PRACTICE OF MEDICINE IN RELATION TO MIND, the Pathology of the Nerve Centres, and the Jurisprudence of Insanity; being a Course of Lectures delivered at Guy's Hospital. Illustrated by Chromolithographic Drawings and Physiological Portraits. 8vo, 14s.

HORACE DOBELL, M.D.
Consulting Physician to the Royal Hospital for Diseases of the Chest, &c.

I.

ON DIET AND REGIMEN IN SICKNESS AND HEALTH, and on the Interdependence and Prevention of Diseases and the Diminution of their Fatality. Seventh edition, 8vo. [*In the Press.*

II.

AFFECTIONS OF THE HEART AND IN ITS NEIGHBOURHOOD. Cases, Aphorisms, and Commentaries. Illustrated by the heliotype process. 8vo, 6s 6d.

JOHN EAGLE.
Member of the Pharmaceutical Society.

A NOTE-BOOK OF SOLUBILITIES. Arranged chiefly for the use of Prescribers and Dispensers. 12mo, 2s 6d.

JOHN ERIC ERICHSEN.
Holme Professor of Clinical Surgery in University College; Senior Surgeon to University College Hospital, &c.

MODERN SURGERY; ITS PROGRESS AND TENDENCIES. Being the Introductory Address delivered at University College at the opening of the Session, 1873-74. Demy 8vo, 1s.

DR. FERBER.

MODEL DIAGRAM OF THE ORGANS IN THE THORAX AND UPPER PART OF THE ABDOMEN. With Letter-press Description. In 4to, coloured, 5s.

AUSTIN FLINT, JR., M.D.

Professor of Physiology and Physiological Anatomy in the Bellevue Medical College, New York; attending Physician to the Bellevue Hospital, &c.

I.

A TEXT-BOOK OF HUMAN PHYSIOLOGY; Designed for the Use of Practitioners and Students of Medicine. Illustrated by plates, and 313 wood engravings, large 8vo, 28s.

II.

THE PHYSIOLOGY OF MAN; Designed to Represent the Existing State of Physiological Science, as applied to the Functions of the Human Body. 5 vols., large 8vo. VOL. I.—The Blood ; Circulation ; Respiration, 18s. VOL. II.— Alimentation; Digestion; Absorption; Lymph and Chyle, 18s. VOL. III.—Secretion; Excretion; Ductless Glands; Nutrition; Animal Heat; Movements; Voice and Speech, 18s. VOL. IV.— The Nervous System, 18s. VOL. V.—Special Senses; Generation, 18s.

J. MILNER FOTHERGILL, M.D.

Member of the Royal College of Physicians of London; Senior Assistant Physician to the West London Hospital, and to the City of London Hospital for Diseases of the Chest, Victoria Park, &c.

I.

THE HEART AND ITS DISEASES, WITH THEIR TREATMENT; INCLUDING THE GOUTY HEART. Second Edition, entirely re-written, copiously illustrated with woodcuts and lithographic plates. 8vo, 16s.

II.

INDIGESTION, BILIOUSNESS, AND GOUT IN ITS PROTEAN ASPECTS.

PART I.—INDIGESTION AND BILIOUSNESS. Post 8vo, 7s 6d.

PART II.—GOUT IN ITS PROTEAN ASPECTS.
[*In Preparation.*

III.

HEART STARVATION. (Reprinted from the Edinburgh Medical Journal), 8vo, 1s.

ERNEST FRANCIS, F.C.S.

Demonstrator of Practical Chemistry, Charing Cross Hospital.

PRACTICAL EXAMPLES IN QUANTITATIVE ANALYSIS, forming a Concise Guide to the Analysis of Water, &c. Illustrated, fcap. 8vo, 2s. 6d.

HENEAGE GIBBES, M.D.

Curator of the Anatomical Museum at King's College.

PRACTICAL HISTOLOGY AND PATHOLOGY.
Crown 8vo, 3s. 6d.

C. A. GORDON, M.D., C.B.

Deputy Inspector General of Hospitals, Army Medical Department.

REMARKS ON ARMY SURGEONS AND THEIR WORKS. Demy 8vo, 5s.

SAMUEL D. GROSS, M.D., LL.D., D.C.L. OXON.

Professor of Surgery in the Jefferson Medical College of Philadelphia.

A PRACTICAL TREATISE ON THE DISEASES, INJURIES, AND MALFORMATIONS OF THE URINARY BLADDER, THE PROSTATE GLAND, AND THE URETHRA. Third Edition, revised and edited by S. W. GROSS, A.M., M.D., Surgeon to the Philadelphia Hospital. Illustrated by 170 engravings, 8vo, 18s.

SAMUEL W. GROSS, A.M., M.D.

Surgeon to, and Lecturer on Clinical Surgery in, the Jefferson Medical College Hospital, and the Philadelphia Hospital, &c.

A PRACTICAL TREATISE ON TUMOURS OF THE MAMMARY GLAND; embracing their Histology, Pathology and Treatment. With Illustrations, 8vo, 10s 6d.

WILLIAM A. HAMMOND, M.D.

Professor of Mental and Nervous Diseases in the Medical Department of the University of the City of New York.

I.

A TREATISE ON THE DISEASES OF THE NERVOUS SYSTEM. Seventh Edition, with 112 Illustrations, large 8vo, 25s.   [*Just Published.*

II.

SPIRITUALISM AND ALLIED CAUSES AND CONDITIONS OF NERVOUS DERANGEMENT. With Illustrations, post 8vo, 8s. 6d.

ALEXANDER HARVEY, M.A., M.D.

Emeritus Professor of Materia Medica in the University of Aberdeen; Consulting Physician to the Aberdeen Royal Infirmary, &c.

FIRST LINES OF THERAPEUTICS; as based on the Modes and processes of Healing, as occurring Spontaneously in Disease; and on the Modes and the Processes of Dying, as resulting Naturally from Disease. In a series of Lectures. Post 8vo, 5s.

ALEXANDER HARVEY, M.D.

Emeritus Professor of Materia Medica in the University of Aberdeen, &c.

AND

ALEXANDER DYCE DAVIDSON, M.D.

Professor of Materia Medica in the University of Aberdeen.

SYLLABUS OF MATERIA MEDICA FOR THE USE OF TEACHERS AND STUDENTS. Based on a selection or definition of subjects in teaching and examining; and also on an estimate of the relative values of articles and preparations in the British Pharmacopœia with doses affixed. Fourth Edition, 16mo, 1s 6d.

GRAILY HEWITT, M.D.

Professor of Midwifery and Diseases of Women in University College, Obstetrical Physician to University College Hospital, &c.

OUTLINES OF PICTORIAL DIAGNOSIS OF DISEASES OF WOMEN. Folio, 6s.

HINTS TO CANDIDATES FOR COMMISSIONS IN THE PUBLIC MEDICAL SERVICES; with Examination Questions, Vocabulary of Hindustani Medical Terms, &c. 8vo, 2s.

F. HOFFMANN, PH.D.

Pharmaceutist in New York.

MANUAL OF CHEMICAL ANALYSIS AS APPLIED TO THE EXAMINATION OF MEDICINAL CHEMICALS. A Guide for the Determination of their Identity and Quality, and for the Detection of Impurities and Adulterations; For the use of Pharmaceutists, Physicians, Druggists, and Manufacturing Chemists, and of Pharmaceutical and Medical Students. With Illustrations, roy. 8vo, 12s.

E. HOLLAND, M.D., F.R.C.S.

HEALTH IN THE NURSERY AND HOW TO FEED AND CLOTHE A CHILD; with Observations on Painless Parturition. A Guide and Companion for the Young Matron and her Nurse. Second Edition, Fcap. 8vo, 2s., paper cover, 1s.

SIR W. JENNER, BART., M.D.

Physician in Ordinary to H.M. the Queen, and to H.R.H. the Prince of Wales.

THE PRACTICAL MEDICINE OF TO-DAY: Two Addresses delivered before the British Medical Association, and the Epidemiological Society. Small 8vo, 1s. 6d.

NORMAN W. KINGSLEY, M.D.S., D.D.S.

President of the Board of Censors of the State of New York; Member of the American Academy of Dental Science, &c.

A TREATISE ON ORAL DEFORMITIES AS A BRANCH OF MECHANICAL SURGERY. With over 350 Illustrations, 8vo, 16s.

E. A. KIRBY, M.D., M.R.C.S.

Late Physician to the City Dispensary.

I.

A FORMULARY OF SELECTED REMEDIES WITH THERAPEUTIC ANNOTATIONS, and a Copious Index of Diseases and Remedies, Diet Tables, &c. A Handbook for Prescribers. Fifth Edition, p. 8vo, 3s 6d.

II.

ON THE VALUE OF PHOSPHORUS AS A REMEDY FOR LOSS OF NERVE POWER. Fifth Edition, 8vo, 2s 6d.

J. WICKHAM LEGG, F.R.C.P.

Assistant Physician to Saint Bartholomew's Hospital, and Lecturer on Pathological Anatomy in the Medical School.

I.

ON THE BILE, JAUNDICE, AND BILIOUS DISEASES. With Illustrations in chromo-lithography, 719 pages, roy. 8vo, 25s.

II.

A GUIDE TO THE EXAMINATION OF THE URINE; intended chiefly for Clinical Clerks and Students. Fifth Edition, revised and enlarged, with additional Illustrations, fcap. 8vo, 2s 6d.

III.

A TREATISE ON HÆMOPHILIA, SOMETIMES CALLED THE HEREDITARY HÆMORRHAGIC DIATHESIS. Fcap. 4to, 7s 6d.

DR. GEORGE LEWIN.

Professor at the Fr. Wilh. University, and Surgeon-in-Chief of the Syphilitic Wards and Skin Disease Wards of the Charité Hospital, Berlin.

THE TREATMENT OF SYPHILIS WITH SUBCU-TANEOUS SUBLIMATE INJECTIONS. Translated by DR. CARL PRŒGLE, and DR. E. H. GALE, late Surgeon United States Army. Small 8vo, 7s.

J. S. LOMBARD, M.D.

Formerly Assistant Professor of Physiology in Harvard College.

I.

EXPERIMENTAL RESEARCHES ON THE RE-GIONAL TEMPERATURE OF THE HEAD, under Conditions of Rest, Intellectual Activity and Emotion. With Illustrations, 8vo, 8s.

II.

ON THE NORMAL TEMPERATURE OF THE HEAD. 8vo, 5s.

WILLIAM THOMPSON LUSK, A.M., M.D.

Professor of Obstetrics and Diseases of Women in the Bellevue Hospital Medical College, &c.

THE SCIENCE AND ART OF MIDWIFERY. With numerous Illustrations, 8vo, 18s.

DR. V. MAGNAN.

Physician to St. Anne Asylum, Paris ; Laureate of the Institute.

ON ALCOHOLISM, THE VARIOUS FORMS OF ALCOHOLIC DELIRIUM AND THEIR TREAT-MENT. Translated by W. S. GREENFIELD, M.D., M.R.C.P. 8vo, 7s. 6d.

PROFESSOR MARTIN.

MARTIN'S ATLAS OF OBSTETRICS AND GYNÆ-COLOGY. Edited by A. MARTIN, Docent in the University of Berlin. Translated and edited with additions by FANCOURT BARNES, M.D., M.R.C.P., Physician to the British Lying-in Hospital ; Assistant Physician to the Royal Maternity Charity of London, &c. Medium 4to, Morocco half bound, 31s. 6d. net.

J. F. MEIGS, M.D.
Consulting Physician to the Children's Hospital, Philadelphia.

AND

W. PEPPER, M.D.
Lecturer on Clinical Medicine in the University of Pennsylvania.

A PRACTICAL TREATISE ON THE DISEASES OF CHILDREN. Seventh Edition, revised and enlarged, roy. 8vo.   [*Nearly Ready.*

DR. MORITZ MEYER.
Royal Counsellor of Health, &c.

ELECTRICITY IN ITS RELATION TO PRACTICAL MEDICINE. Translated from the Third German Edition, with notes and additions by WILLIAM A. HAMMOND, M.D. With Illustrations, large 8vo, 18s.

WM. JULIUS MICKLE, M.D., M.R.C.P.
Member of the Medico-Psychological Association of Great Britain and Ireland ; Member of the Clinical Society, London ; Medical Superintendent, Grove Hall Asylum, London.

GENERAL PARALYSIS OF THE INSANE. 8vo, 10s.

E. A. MORSHEAD, M.R.C.S., L.R.C.P.
Assistant to the Professor of Medicine in University College, London.

TABLES OF THE PHYSIOLOGICAL ACTION OF DRUGS. Fcap. 8vo, 1s.

A. STANFORD MORTON, M.B., F.R.C.S. ED.
Senior Assistant Surgeon, Royal South London Ophthalmic Hospital.

REFRACTION OF THE EYE: Its Diagnosis, and the Correction of its Errors, with Chapter on Keratoscopy. Small 8vo, 2s. 6d.

WILLIAM MURRELL, M.D. M.R.C.P., M.R.C.S.
Lecturer on Materia Medica and Therapeutics at Westminster Hospital ; Senior Assistant Physician, Royal Hospital for Diseases of the Chest.

I.

WHAT TO DO IN CASES OF POISONING. Med 32mo, 1s. 6d.

II.

NITRO-GLYCERINE AS A REMEDY FOR ANGINA PECTORIS. Crown 8vo.   [*Just ready.*

WILLIAM NEWMAN, M.D., F.R.C.S.

Surgeon to the Stamford Infirmary.

SURGICAL CASES: Mainly from the Wards of the Stamford, Rutland, and General Infirmary. 8vo, paper boards, 4s. 6d.

DR. FELIX VON NIEMEYER.

Late Professor of Pathology and Therapeutics ; Director of the Medical Clinic of the University of Tübingen.

A TEXT-BOOK OF PRACTICAL MEDICINE, WITH PARTICULAR REFERENCE TO PHYSIOLOGY AND PATHOLOGICAL ANATOMY. Translated from the Eighth German Edition, by special permission of the Author, by GEORGE H. HUMPHREY, M.D., and CHARLES E. HACKLEY, M.D., Revised Edition, 2 vols., large 8vo, 36s.

C. F. OLDHAM, M.R.C.S., L.R.C.P.

Surgeon H.M. Indian Forces ; late in Medical charge of the Dalhousie Sanitarium.

WHAT IS MALARIA? AND WHY IS IT MOST INTENSE IN HOT CLIMATES? An explanation of the Nature and Cause of the so-called Marsh Poison, with the Principles to be observed for the Preservation of Health in Tropical Climates and Malarious Districts. Demy 8vo, 7s. 6d.

G. OLIVER, M.D., M.R.C.P.

THE HARROGATE WATERS: Data Chemical and Therapeutical, with notes on the Climate of Harrogate. Addressed to the Medical Profession. Crown 8vo, with Map of the Wells, 3s. 6d.

JOHN S. PARRY, M.D.

Obstetrician to the Philadelphia Hospital, Vice-President of the Obstetrical and Pathological Societies of Philadelphia, &c.

EXTRA-UTERINE PREGNANCY; Its Causes, Species, Pathological Anatomy, Clinical History, Diagnosis, Prognosis and Treatment. 8vo, 8s.

E. RANDOLPH PEASLEE, M.D., LL.D.

Late Professor of Gynæcology in the Medical Department of Dartmouth College;
President of the New York Academy of Medicine, &c.,&.

OVARIAN TUMOURS: Their Pathology, Diagnosis, and
Treatment, especially by Ovariotomy. Illustrations, roy.
8vo, 16s.

R. DOUGLAS POWELL, M.D., F.R.C.P.

Physician to the Hospital for Consumption and Diseases of the Chest at
Brompton, Physician to the Middlesex Hospital.

ON CONSUMPTION AND ON CERTAIN DISEASES
OF THE LUNGS AND PLEURA. Being a Second
Edition revised and extended of "The Varieties of Pulmonary
Consumption." Illustrated by woodcuts and a coloured plate,
8vo, 9s.

AMBROSE L. RANNEY, A.M., M.D.

Adjunct Professor of Anatomy in the University of New York, etc.

THE APPLIED ANATOMY OF THE NERVOUS
SYSTEM, being a study of this portion of the Human
Body from a stand-point of its general interest and practical
utility, designed for use as a Text-book and a work of Reference.
With 179 Illustrations, 8vo, 20s.   [*Now Ready*.

RALPH RICHARDSON, M.A., M.D.

Fellow of the College of Physicians Edinburgh.

ON THE NATURE OF LIFE: An Introductory Chapter
to Pathology. Second Edition, revised and enlarged.
Fcap. 4to 10s. 6d.

W. RICHARDSON, M.A., M.D., M.R.C.P.

REMARKS ON DIABETES, ESPECIALLY IN RE-
FERENCE TO TREATMENT. Demy 8vo, 4s. 6d.

SYDNEY RINGER, M.D.,

Professor of the Principles and Practice of Medicine in University College;
Physician to and Professor of Clinical Medicine in, University
College Hospital.

I.

A HANDBOOK OF THERAPEUTICS: Ninth Edition,
8vo.   [*Just Ready*.

II.

ON THE TEMPERATURE OF THE BODY AS A
MEANS OF DIAGNOSIS AND PROGNOSIS IN
PHTHISIS. Second Edition, small 8vo, 2s. 6d.

FREDERICK T. ROBERTS, M.D., B.SC., F.R.C.P.

Examiner in Medicine at the Royal College of Surgeons ; Professor of Thera-
peutics in University College ; Physician to, and Professor of Clinical
Medicine in, University College Hospital ; Physician to
the Brompton Consumption Hospital, &c.

A HANDBOOK OF THE THEORY AND PRACTICE OF MEDICINE. Fourth Edition, with Illustrations, 2 vols., 8vo, 22s.

D. B. ST. JOHN ROOSA, M.A., M.D.

Professor of Diseases of the Eye and Ear in the University of the City of New
York ; Surgeon to the Manhattan Eye and Ear Hospital ; Consulting Sur-
geon to the Brooklyn Eye and Ear Hospital. &c.,&c.

A PRACTICAL TREATISE ON THE DISEASES OF THE EAR, including the Anatomy of the Organ. Fourth Edition, Illustrated by wood engravings and chromo-lithographs large 8vo, 22s.

J. BURDON SANDERSON, M.D., LL.D., F.R.S.

Jodrell Professor of Physiology in University College, London.

I.

SYLLABUS OF A COURSE OF LECTURES ON PHYSIOLOGY. Second Edition, 8vo, 4s.

II.

UNIVERSITY COLLEGE COURSE OF PRACTICAL EXERCISES IN PHYSIOLOGY. With the co-operation F. J. M. PAGE, B.Sc. F.C.S.; W. NORTH, B.A., F.C.S., and AUG. WALLER, M.D. Demy 8vo, 3s. 6d. [*Just published.*

ALDER SMITH, M.B. LOND., F.R.C.S.

Resident Medical Officer, Christ's Hospital, London.

RINGWORM: ITS DIAGNOSIS AND TREATMENT. Second Edition, rewritten and enlarged. With additional Illustrations, fcap. 8vo. [*Just ready.*

J. LEWIS SMITH, M.D.

Physician to the New York Infants' Hospital ; Clinical Lecturer on Diseases
of Children in Bellevue Hospital Medical College.

A TREATISE ON THE DISEASES OF INFANCY AND CHILDHOOD. Fifth Edition, with Illustrations, large 8vo, 21s. [*Now Ready.*

JAMES STARTIN, M.B., M.R.C.S.

Surgeon and Joint Lecturer to St. John's Hospital for Diseases of the Skin.

LECTURES ON THE PARASITIC DISEASES OF THE SKIN. Vegetoid and Animal. With Illustrations Crown 8vo, 3s. 6d.

LEWIS A. STIMSON, B.A., M.D.

Surgeon to the Presbyterian Hospital; Professor of Pathological Anatomy in the Medical Faculty of the University of the City of New York.

A MANUAL OF OPERATIVE SURGERY. With three hundred and thirty-two Illustrations. Post 8vo, 10s. 6d.

HUGH OWEN THOMAS, M.R.C.S.

I.

DISEASES OF THE HIP, KNEE, AND ANKLE JOINTS, with their Deformities, treated by a new and efficient method. With an Introduction by RUSTHON PARKER, F.R.C.S., Lecturer on Surgery at the School of Medicine, Liverpool. Third Edition, 8vo, 25s.

II.

THE PAST AND PRESENT TREATMENT OF IN-TESTINAL OBSTRUCTIONS, reviewed, with an Improved Treatment Indicated. Second Edition, 8vo, 10s. 6d.

J. C. THOROWGOOD, M.D.

Assistant Physician to the City of London Hospital for Diseases of the Chest.

THE CLIMATIC TREATMENT OF CONSUMPTION AND CHRONIC LUNG DISEASES. Third Edition, post 8vo, 3s. 6d.

J. ASHBURTON THOMPSON, M.R.C.S.

FREE PHOSPHORUS IN MEDICINE WITH SPE-CIAL REFERENCE TO ITS USE IN NEURALGIA, A contribution to Materia Medica and Therapeutics. An account of the History, Pharmaceutical Preparations, Dose, Internal Administration, and Therapeutic uses of Phosphorus; with a Complete Bibliography of this subject, referring to nearly 200 works upon it. Demy 8vo, 7s. 6d.

LAURENCE TURNBULL, M.D., PH.G.

Aural Surgeon to Jefferson Medical College Hospital, &c., &c.

ARTIFICIAL ANÆSTHESIA : A Manual of Anæsthetic Agents, and their Employment in the Treatment of Disease. Second Edition, with Illustrations, crown 8vo, 6s.

W. H. VAN BUREN, M.D. LL.D.

Professor of Surgery in the Bellevue Hospital Medical College.

Diseases of the Rectum : and the Surgery of the Lower Bowel. Second Edition, with Illustrations, 8vo, 14s.   [*Just Published*.

RUDOLPH VIRCHOW, M.D.

Professor in the University, and Member of the Academy of Sciences of Berlin, &c.

Infection-Diseases in the Army, chiefly Wound Fever, Typhoid, Dysentery, and Diphtheria. Translated from the German by John James, M.B., F.R.C.S. Fcap. 8vo, 1s. 6d.

ALFRED VOGEL, M.D.

Professor of Clinical Medicine in the University of Dorpat, Russia.

A Practical Treatise on the Diseases of Children. Translated and Edited by H. Raphael, M.D. From the Fourth German Edition, Illustrated by six lithographic plates, part coloured, large 8vo, 18s.

A. DE WATTEVILLE, M.A. B.SC., M.R.C.S.

Assistant Physician to the Hospital for Epilepsy and Paralysis, late Electro-Therapeutical Assistant to University College, Hospital.

A Pratical Introduction to Medical Electricity. Second Edition.   [*In Preparation*.

W. SPENCER WATSON, F.R.C.S., M.B.

Surgeon to the Great Northern Hospital ; Surgeon to the Royal South London Ophthalmic Hospital.

I.

Eyeball Tension : its Effects on the Sight and its Treatment. With woodcuts, p. 8vo, 2s. 6d.

II.

Diseases of the Nose and its Accessory Cavities. Profusely Illustrated. Demy 8vo, 18s.

III.

On Abscess and Tumours of the Orbit. p. 8vo, 2s. 6d.

DR. F. WINCKEL

Formerly Professor and Director of the Gynæcological Clinic at the University of Rostock.

THE PATHOLOGY AND TREATMENT OF CHILD-BED: A Treatise for Physicians and Students. Translated from the Second German Edition, with many additional notes by the Author, by J. R. CHADWICK, M.D. 8vo 14s.

EDWARD WOAKES, M.D.

Surgeon to the Ear Department of the Hospital for Diseases of the Throat and Chest ; and Surgeon to the Hospital.

ON DEAFNESS, GIDDINESS AND NOISES IN THE HEAD. Second Edition, with Illustrations, crown 8vo, 7s.

E. T. WILSON, B.M. OXON., F.R.C.P. LOND.

Physician to the Cheltenham General Hospital and Dispensary.

DISINFECTANTS AND HOW TO USE THEM. In Packets of one doz. price 1s.

CLINICAL CHARTS FOR TEMPERATURE OBSERVATIONS, ETC.

Arranged by W. RIGDEN, M.R.C.S. Price 7s. per 100, or 1s. per doz.

Each Chart is arranged for four weeks, and is ruled at the back for making notes of cases; they are convenient in size, and are suitable both for hospital and private practice.

*** MR. LEWIS has transactions with the leading publishing firms in America for the sale of his publications in that country. Arrangements are made in the interests of Authors either for sending a number of copies of their works to the United States, or having them reprinted there, as may be most advantageous.

Mr. Lewis's publications can be procured of any bookseller in any part of the world.

List of works issued for the New Sydenham Society on application.

London : Printed by H. K. Lewis, Gower Street. W.C.